Solutions and Tests For Exploring Creation With Physical Science

Manufactured in the United States of America
Seventh Printing 2005

Published By

Apologia Educational Ministries, Inc.

Anderson, IN

Printed by

The C.J. Krehbiel Company

Cincinnati, OH

Exploring Creation With Physical Science

Solutions and Tests

TABLE OF CONTENTS

Teacher's Notes 1

Solutions to the Study Guide

Solutions to the Study Guide for Module #1 6
Solutions to the Study Guide for Module #2 9
Solutions to the Study Guide for Module #3 11
Solutions to the Study Guide for Module #4 14
Solutions to the Study Guide for Module #5 16
Solutions to the Study Guide for Module #6 18
Solutions to the Study Guide for Module #7 21
Solutions to the Study Guide for Module #8 23
Solutions to the Study Guide for Module #9 25
Solutions to the Study Guide for Module #10 29
Solutions to the Study Guide for Module #11 32
Solutions to the Study Guide for Module #12 35
Solutions to the Study Guide for Module #13 38
Solutions to the Study Guide for Module #14 41
Solutions to the Study Guide for Module #15 45
Solutions to the Study Guide for Module #16 48

Tests

Test for Module #1 53
Test for Module #2 55
Test for Module #3 57
Test for Module #4 59
Test for Module #5 61
Test for Module #6 63
Test for Module #7 65
Test for Module #8 67
Test for Module #9 69
Test for Module #10 71
Test for Module #11 73
Test for Module #12 75
Test for Module #13 77
Test for Module #14 79
Test for Module #15 81
Test for Module #16 83

Answers to the Tests

Answers to the Module #1 Test 86
Answers to the Module #2 Test 88
Answers to the Module #3 Test 90
Answers to the Module #4 Test 92
Answers to the Module #5 Test 93
Answers to the Module #6 Test 95
Answers to the Module #7 Test 96
Answers to the Module #8 Test 97
Answers to the Module #9 Test 98
Answers to the Module #10 Test 101
Answers to the Module #11 Test 103
Answers to the Module #12 Test 104
Answers to the Module #13 Test 106
Answers to the Module #14 Test 108
Answers to the Module #15 Test 111
Answers to the Module #16 Test 113

TEACHER'S NOTES
Exploring Creation With Physical Science
Dr. Jay L. Wile

Thank you for purchasing *Exploring Creation With Physical Science*. I designed this modular course specifically to meet the needs of the homeschooling parent. I am very sensitive to the fact that most homeschooling parents do not know the upper-level sciences very well, if at all. As a result, they consider it nearly impossible to teach to their children. This course has several features that make it ideal for such a parent.

1. The course is written in a conversational style. Unlike many authors, I do not get wrapped up in the desire to write formally. As a result, the text is easy to read and the student feels more like he or she is *learning*, not just reading.

2. The course is completely self-contained. Each module includes the text of the lesson, experiments to perform, questions to answer, and a test to take. The solutions to the questions are fully explained, and the test answers are provided. The experiments are written in a careful, step-by-step manner that tells the student not only what he or she should be doing, but also what he or she should be observing.

3. Most importantly, this course is Christ-centered. In every way possible, I try to make physical science glorify God. One of the most important things that you and your student should get out of this course is a deeper appreciation for the wonder of God's Creation!

I hope that you enjoy using this course as much as I enjoyed writing it!

Pedagogy of the Text

(1) There are two types of exercises that the student is expected to complete: "on your own" questions, and an end-of-the module study guide.

- The "on your own" questions should be answered as the student reads the text. The act of answering these questions will cement in the student's mind the concepts he or she is trying to learn. Answers to these problems are in the student text.

- The study guide should be completed in its entirety after the student has finished each module. Answers to the study guide questions are in this book.

The student should be allowed to study the solutions to the "on your own" questions while he or she is working on them. When the student reaches the study guide, however, the solutions should be used only to check the student's completed work.

(2) In addition to the solutions to the study guides, there is a test for each module in this book, along with the answers to the test. **I strongly recommend that you administer**

**each test once the student has completed the module and all associated exercises.
The student should be allowed to have only pencil, paper, and any tables that are
specifically mentioned in the test.** I understand that many homeschoolers do not like the
idea of administering tests. However, if your student is planning to attend college, it is
absolutely necessary that he or she become comfortable with taking tests!

(3) The best way to grade the tests is to assign one point for every answer that the student
must supply. Thus, if a question has three parts and an answer must be supplied for each
part, the question should be worth 3 points. The student's percentage correct, then, is
simply the number of answers the student got right divided by the total number of
answers times 100. The student's letter grade should be based on a 90/80/70/60 scale.

(4) All definitions presented in the text are centered. The words will appear in the study
guide and their definitions need to be memorized.

(5) Words that appear in bold-face type in the text are important terms that the student
should know.

(6) The study guide gives your student a good feel for what I require him or her to know
for the test. Any information needed to answer the study guide questions is information
that the student must know for the test. Sometimes, tables and other reference material will
be provided on a test so that the student need not memorize it. You will be able to tell if
this is the case because the questions in the study guide which refer to this information will
specifically tell the student that he or she can use the reference material.

<u>Experiments</u>

The experiments in this course are designed to be done as the student is reading
the text. I recommend that your student keep a notebook of these experiments. This
notebook serves two purposes. First, as the student writes about the experiment in the
notebook, he or she will be forced to think through all of the concepts that were explored
in the experiment. This will help the student cement them into his or her mind. Second,
certain colleges might actually ask for some evidence that your student did, indeed, have
a laboratory component to his or her physical science course. The notebook will not only
provide such evidence but will also show the college administrator the quality of the
physical science instruction that you provided to your student. I recommend that you
perform your experiments in the following way:

- When your student gets to the experiment during the reading, have him or her read
  through the experiment in its entirety. This will allow the student to gain a quick
  understanding of what her or she is to do.

- Once the student has read the experiment, he or she should then start a new page in
  his or her laboratory notebook. The first page should be used to write down all of the

data taken during the experiment and perform any exercises discussed in the experiment.

- When the student has finished the experiment, he or she should write a brief report in his or her notebook, right after the page where the data and exercises were written. The report should be a brief discussion of what was done and what was learned.

- **PLEASE OBSERVE COMMON SENSE SAFETY PRECAUTIONS. The experiments are no more dangerous than most normal, household activity. Remember, however, that the vast majority of accidents do happen in the home!**

Question/Answer Service

For all those who use my curriculum, I offer a question/answer service. If there is anything in the modules that you do not understand - from an esoteric concept to a solution for one of the problems - just get in touch with me by any of the means listed on the **NEED HELP?** page that is in the student textbook.

Solutions To The

Study
Guides

SOLUTIONS TO THE MODULE #1 STUDY GUIDE

1. a. <u>Atom</u> - The smallest stable unit of matter in Creation

b. <u>Molecule</u> - Two or more atoms linked together to make a substance with unique properties

c. <u>Concentration</u> - The quantity of a substance within a certain volume of space

2. <u>Carbon disulfide must be a molecule</u>. Since it can be broken down into smaller components, then it must be made of molecules. If it were made of atoms, then it could not be broken down into smaller, stable units.

3. Rust is not attracted to a magnet because <u>when an atom is a part of a molecule, the molecule does not take on the characteristics of the atom. Instead, the atoms in the molecule join together in such a way as to give the molecule its unique characteristics.</u>

4. <u>The statue will eventually turn a shade of green, just like the copper wire did in Experiment 1.1.</u> This comes from the copper atoms reacting with water and carbon dioxide in the air to make copper hydroxycarbonate.

5. <u>Scientists have NOT seen atoms</u>. The scanning tunneling electron microscope "pictures" that you see are not pictures of atoms. Instead, they are the result of computer calculations involving electricity and a theory call quantum mechanics.

6. <u>Kilo means 1,000; centi means 0.01; and milli means 0.001</u>.

7. <u>Mass is measured in grams in the metric system. In the English system, it is measured in slugs</u>.

8. <u>Volume is measured in liters in the metric system. In the English system, it is measured in gallons, pints, or quarts</u>.

9. <u>Length is measured in meters in the metric system. In the English system, it is measured in feet, yards, inches, or miles</u>.

10. First, we convert the number to a fractional form:

$$\frac{1.3 \text{ m}}{1}$$

Next, since we want to convert from meters to centimeters, we need to remember that "centi" means "0.01." So one centimeter is the same thing as 0.01 meters. Thus:

$$1 \text{ cm} = 0.01 \text{ m}$$

That's our conversion relationship. Since we want to end up with cm in the end, then we must multiply the measurement by a fraction that has meters on the bottom (to cancel the meter unit that is there) and cm on the top (so that cm is the unit we are left with). Remember, the numbers next to the units in the relationship above go with the units. Thus, since "m" goes on the bottom of the fraction, so does "0.01." Since "cm" goes on the top, so does "1."

$$\frac{1.3 \text{ m}}{1} \times \frac{1 \text{ cm}}{0.01 \text{ m}} = 130 \text{ cm}$$

Therefore, 1.3 m is the same as <u>130 cm</u>.

11. First, we convert the number to a fractional form:

$$\frac{75 \text{ kg}}{1}$$

Next, since we want to convert from kg to grams, we need to remember that "kilo" means "1,000." So one kilogram is the same thing as 1,000 grams. Thus:

$$1 \text{ kg} = 1,000 \text{ g}$$

That's our conversion relationship. Since we want to end up with grams in the end, then we must multiply the measurement by a fraction that has kilograms on the bottom (to cancel the kg unit that is there) and grams on the top (so that g is the unit we are left with):

$$\frac{75 \text{ kg}}{1} \times \frac{1,000 \text{ g}}{1 \text{ kg}} = 75,000 \text{ g}$$

The person's mass is <u>75,000 g</u>.

12. We use the same procedure that we used in the previous two problems. Thus, I am going to reduce the length of the explanation.

$$\frac{0.5 \text{ gal}}{1} \times \frac{3.78 \text{ L}}{1 \text{ gal}} = 1.89 \text{ L}$$

There are <u>1.89 L</u> in half a gallon.

13.
$$\frac{100 \text{ cm}}{1} \times \frac{1 \text{ inch}}{2.54 \text{ cm}} = 39.37 \text{ in}$$

There are <u>39.37 inches</u> in a meterstick. Note that I rounded the answer. The real answer was "39.370078740," but there are simply too many digits in that number. When you take chemistry, you will learn about significant figures, a concept that tells you where to round numbers off. For right now, don't worry about it. If you rounded at a different spot than I did, that's fine.

14. <u>Baking bread is not a dangerous activity because the ozone it produces is not concentrated enough to be dangerous</u>. Ozone is a poison, but at low enough concentrations, it does not adversely affect people. At higher concentrations, however, it can be toxic enough to kill you!

SOLUTIONS TO THE MODULE #2 STUDY GUIDE

1. a. <u>Humidity</u> - The moisture content of air

b. <u>Absolute humidity</u> - The mass of water vapor contained in a certain volume of air

c. <u>Relative humidity</u> - A quantity expressing humidity as a percentage of the maximum absolute humidity for that particular temperature

d. <u>Greenhouse effect</u> - The process by which certain gases (principally water, carbon dioxide, and methane) trap heat that would otherwise escape the earth and radiate into space

e. <u>Parts per million</u> - The number of molecules (or atoms) of a substance in a mixture for every one million molecules (or atoms) in that mixture

2. <u>The humidity is higher on the first day.</u> Since the person felt cooler on the second day (despite the same temperature), his sweat must have evaporated more quickly than on the first day. Thus, the first day was more humid.

3. <u>The child will add more water on the second day.</u> Since the humidity was lower on the second day, the water in the bowl will evaporate more quickly.

4. <u>The water will not evaporate.</u> Since the relative humidity is 100%, the air cannot hold any more water vapor. As a result, no water will evaporate from the glass.

5. <u>Sweat cools you off because when it evaporates, it takes energy from your skin.</u> When energy leaves your skin, it gets cooler.

6. <u>Dry air is 78% nitrogen and 21% oxygen.</u>

7. <u>If the air had no carbon dioxide in it, the earth would be colder.</u> Since carbon dioxide is a greenhouse gas, the greenhouse effect would be weaker, leaving a cool earth. You could also answer this question by saying that plants would die of starvation.

8. <u>If there were no ozone in the air, ultraviolet light would kill all living things.</u>

9. <u>If more oxygen were in the air, lifespans would decrease and forest fires would increase in frequency and ferocity.</u>

10. <u>There is no reason to expect that the new planet will have the same temperature as earth. If it does not have essentially the same air, with all the same levels of all the greenhouse gases, then it will not have the same temperature!</u>

11. <u>Nitrogen makes up the majority of the air we exhale.</u> See Figure 2.3.

12. <u>We exhale more oxygen.</u> See Figure 2.3.

13. <u>No.</u> Figure 2.4 shows that the average global temperature has been reasonably constant for the past 70 years.

14. Remember, we know the relationship between ppm and percent. We can therefore just use the factor-label method to figure out the answer.

$$\frac{0.110 \; \cancel{ppm}}{1} \times \frac{1\%}{10,000 \; \cancel{ppm}} = 0.0000110\%$$

A concentration of 11 ppm is equal to <u>0.000011%</u>.

15. Remember, we know the relationship between percent and ppm, so we can convert using the factor-label method.

$$\frac{0.023 \; \cancel{\%}}{1} \times \frac{10,000 \; ppm}{1 \cancel{\%}} = 230 \; ppm$$

The concentration of nitrogen oxides in this sample of air is <u>230 ppm</u>.

16. <u>The air is much cleaner today than 20 years ago.</u> See Figure 2.6.

17. <u>A cost/benefit analysis attempts to determine whether or not to take an action by determining the benefits of that action as well as the costs. If the benefit outweighs the cost, then the action should be taken. If not, the action should not be taken.</u>

18. <u>A catalytic converter converts carbon monoxide in the car's exhaust to carbon dioxide.</u> The fact that most cars have these today is responsible for cutting in half the carbon monoxide concentration in the air.

19. <u>A scrubber traps sulfur oxides in a smokestack and keeps them from being emitted into the air.</u> These have been largely responsible for the 70% decrease in sulfur oxides concentration in the air.

20. <u>Ground-level ozone is a pollutant because it is a poison, and it is where we can breathe it. Ozone in the ozone layer is not a pollutant because no one breathes that high up in the air, so its poisonous properties are unimportant. It is necessary in the ozone layer in order to block the sun's ultraviolet rays.</u>

SOLUTIONS TO THE MODULE #3 STUDY GUIDE

1. a. <u>Atmosphere</u> - The mass of air surrounding a planet

b. <u>Atmospheric pressure</u> - The pressure exerted by the atmosphere on all objects within it

c. <u>Barometer</u> - An instrument used to measure atmospheric pressure

d. <u>Homosphere</u> - The lower layer of earth's atmosphere, which exists from ground level to roughly 80 kilometers (50 miles) above sea level

e. <u>Heterosphere</u> - The upper layer of earth's atmosphere, which exists higher than 80 kilometers (50 miles) above sea level

f. <u>Troposphere</u> - The region of the atmosphere that extends from ground level to roughly 11 kilometers (7 miles) above sea level

g. <u>Stratosphere</u> - The region of the atmosphere that spans altitudes of 11 kilometers to 48 kilometers (30 miles)

h. <u>Mesosphere</u> - The region of the atmosphere that spans altitudes of 48 kilometers to 80 kilometers (50 miles)

i. <u>Jet streams</u> - Narrow bands of high-speed winds that circle the earth, blowing from west to east

j. <u>Heat</u> - Energy that is being transferred

k. <u>Temperature</u> - A measure of the energy of motion in a substance's molecules

l. <u>Thermosphere</u> - The region of the atmosphere between altitudes of 80 kilometers and 460 kilometers

m. <u>Exosphere</u> - The region of the atmosphere above an altitude of 460 kilometers

n. <u>Ionosphere</u> - The region of the atmosphere between the altitudes of 65 kilometers and 330 kilometers where the gases are ionized

2. <u>Atmospheric pressure would be greater than it is now.</u> After all, if the air supply were twice as concentrated, there would be twice as much air. Thus, the mass of air pressing down on everything in the atmosphere would be twice as high.

3. <u>The second student's barometer (the one made with water) will have a much higher column.</u> Remember how a barometer works. The height of the column of liquid is determined by the amount of liquid necessary to balance out the atmospheric pressure pushing on the liquid outside of the column. The heavier the liquid, the less will be necessary to achieve the effect. Since

mercury is heavier (by volume) than water, then it will take less mercury to balance out atmospheric pressure. Thus, the mercury column will be smaller than the water column.

4. <u>An atmospheric pressure of 25.4 inches of mercury would be reported</u>. Since 1.0 atm corresponds to the average sea-level value of atmospheric pressure, then 0.85 atms means that the atmospheric pressure is lower than average.

5. <u>The first came from the homosphere</u>. In the homosphere, the mixture of gases in the air is the same throughout. It is the mixture we learned in the previous module. The heterosphere has many different compositions, depending on altitude.

6. <u>The balloon enters the stratosphere when its temperature readings cease to decrease and begin increasing. The balloon enters the mesosphere when the temperature readings cease increasing and begin decreasing again</u>. Since the temperature gradient changes at the stratosphere and then again at the mesosphere, this can be used to determine when the balloon has reached those parts of the atmosphere.

7. <u>Troposphere, stratosphere, mesosphere</u>

8. <u>I am referring to the "concentration gradient." You could also answer with "pressure gradient."</u> Whether you use the term concentration or pressure, both quantities continue to decrease with increasing altitude. Remember, "gradient" just means steady change, so I can use that term with any quantity.

9. <u>The plane is traveling in the troposphere</u>. That's where the majority of weather phenomena exist.

10. <u>The second vial contains the gas with the highest temperature</u>. Remember, temperature is a measure of the motional energy of a substance. Since the molecules in the second vial have a higher speed, they have more motional energy and thus a higher temperature.

11. <u>Your companion is correct. Heat is energy that is being transferred. The reason you are cold is that energy is being transferred from your body to the surrounding air</u>. Even though it sounds weird to say it, you get cold because of transferred energy; thus, you get cold because of the heat!

12. <u>The temperature gradient would reverse, getting warmer near that region</u>. Remember, the temperature increases with increasing altitude in the stratosphere because of a layer of the greenhouse gas ozone. Carbon dioxide is also a greenhouse gas, and thus would produce roughly the same effect.

13. <u>A ban on CFCs will not save lives because CFCs cause a depletion of ozone only during a few months out of the year and only over Antarctica</u>. Since there is no significant population there, and since the depletion is temporary, the "ozone hole" is not a big threat to human survival.

14. A ban on CFCs will cost many lives because refrigeration, surgical sterilization, and firefighting will all be less efficient, causing death by starvation, death by eating spoiled food, death by surgical infection, and death by fire.

15. The kinds of human-made molecules that contain chlorine can survive the trip up to the ozone layer, while most naturally-produced chlorine-containing molecules cannot. Thus, although we produce few chlorine-containing molecules, almost all of them can reach the ozone layer, where ozone depletion can occur. As a result, most of the ozone-destroying molecules in the ozone layer are from human sources.

16. The Polar Vortex lifts the CFCs into the ozone layer. Since the Polar Vortex is seasonal and limited to the South Pole, so is ozone depletion.

17. The ionosphere is a stretch of the atmosphere ranging from the upper mesosphere to the lower parts of the thermosphere. It is useful to us in radio communication, as radio signals can bounce off of it to extend their range. An altitude range of 65 km to 330 km is also a valid answer to where the ionosphere is.

SOLUTIONS TO THE MODULE #4 STUDY GUIDE

1. a. <u>Electrolysis</u> - Using electricity to break a molecule down into its constituent elements

b. <u>Polar molecule</u> - A molecule that has slight positive and negative charges due to an imbalance in the way electrons are shared

c. <u>Solvent</u> - A liquid substance capable of dissolving other substances

d. <u>Solute</u> - A substance that is dissolved in a solvent

e. <u>Cohesion</u> - The phenomenon that occurs when individual molecules are so strongly attracted to each other that they tend to stay together, even when exposed to tension

f. <u>Hard water</u> - Water that has certain dissolved ions in it, predominately calcium ions

2. <u>The result would be (b) equal amounts of oxygen and hydrogen</u>. The chemical formula says that there are 2 hydrogens and 2 oxygens in each molecule of hydrogen peroxide. Thus, there are equal amounts of each atom, resulting in equal amounts of each gas.

3. <u>The chemical formula HO would be the more likely erroneous result</u>. If the test tube which held hydrogen had a slow leak, then it would look like less hydrogen was collected than what should have been collected. Thus, the experiment would indicate a chemical formula with *less* hydrogen atoms in it. H_4O is an erroneous result that indicates there were *more* hydrogens, HO is the one that indicates *less* hydrogens.

4. There are no subscripts after the Mg or the S, indicating <u>one magnesium atom and one sulfur atom</u>. There is a subscript of 4 after the O, indicating <u>four oxygen atoms</u>.

5. You put the subscripts after each symbol to indicate the number of atoms. If the number is one, there is no subscript. This leads to an answer of <u>$CaCO_3$</u>.

6. There is no subscript after the N, indicating one atom there. The next letter is capital, so it must represent another atom. There is a subscript of 3 after it, indicating three of those. Thus, there are a total of <u>4</u> atoms.

7. <u>The molecule will be nonpolar</u>. If the atoms all pull on electrons with the same strength, none will be able to get more than its fair share.

8. <u>Baking soda will not dissolve in vegetable oil</u>. Since baking soda dissolved in water, it is either ionic or polar (it is ionic). Either way, it will not dissolve in a nonpolar substance because only other nonpolar substances will dissolve in nonpolar substances.

9. <u>Carbon tetrachloride must be nonpolar</u>, otherwise it would have dissolved in water.

10. <u>The liquid would have more molecules</u> because the molecules in solid water are farther apart than they are in liquid water. For the same volume, then, there will be more molecules in the liquid.

11. <u>For any other substance, the answer would be that the solid would have more molecules</u>. For virtually any other substance, molecules are closer together in the solid state, so there would be more molecules in an equal volume as compared to the liquid.

12. <u>Hydrogen bonding is responsible</u>. Hydrogen bonding brings the molecules close together and makes them want to stay close together.

13. <u>Cohesion causes surface tension</u>. You probably could say hydrogen bonding here if you are talking about water. However, other substances exhibit surface tension, even if they do not hydrogen bond. Hydrogen bonding just makes water's surface tension larger than many other substances.

14. <u>Water is harder in some regions of the world because there is a higher amount of calcium-containing substances in some regions than others</u>. It is these dissolved calcium-containing substances that cause hard water.

SOLUTIONS TO THE MODULE #5 STUDY GUIDE

1. a. Hydrosphere - The mass of water on a planet

b. Hydrologic cycle - The process by which water is continuously exchanged between earth's various water sources

c. Transpiration - Emission of water vapor from plants

d. Condensation - The process by which water vapor turns into liquid water

e. Precipitation - Water falling from the atmosphere as rain, snow, sleet, or hail

f. Distillation - Evaporation and condensation of a mixture to separate out the mixture's individual components

g. Residence time - The average time a given molecule of water will stay in a given water source

h. Salinity - A measure of the quantity of dissolved salt in water

i. Firn - A dense, icy pack of snow

j. Water table - The imaginary line between the water-saturated soil and the soil not saturated with water

k. Percolation - The process by which water passes from above the water table to below it

l. Adiabatic cooling - The cooling of a gas that happens when the gas expands

m. Cloud condensation nuclei - Small particles that water vapor condenses on to form clouds

2. The vast majority of water on the earth is saltwater, since more than 97% of earth's water supply is in the oceans.

3. The largest source of freshwater are the glaciers and icebergs on the planet.

4. The largest source of liquid freshwater is groundwater.

5. Water can enter the atmosphere through evaporation and transpiration.

6. If the raindrop never really soaks into the soil, it can end up in a river via surface runoff. It could also soak into the groundwater and get to the river via groundwater flow. Alternatively, it could go into the soil, be absorbed by a plant, transpired into the atmosphere, condensed into a cloud, and precipitated into the river. It could also evaporate before it soaks into the ground. That's four answers, but you only need three of them.

7. <u>Transpiration</u> takes water from the soil because plants absorb the soil moisture and then put it into the atmosphere via transpiration.

8. <u>The residence time in the river is shorter</u>. The residence time will be shorter wherever water is exchanged with other sources quickly.

9. <u>A lake must have a way to get rid of water other than just evaporation</u>. This usually is accomplished when the lake feeds a river or stream. If evaporation is the only way of getting rid of water, then the salts that the lake receives will become concentrated, making saltwater.

10. <u>The oceans are not salty enough for the earth to be billions of years old</u>. Since salt accumulates in the oceans, the older the earth is, the saltier the oceans will be. Calculations indicate that even assuming the oceans had no salt to begin with, it would take only 1 million years (*not billions of years!*) to make the oceans as salty as they are now.

11. <u>Melted sea ice would taste salty</u> because salt is incorporated into sea ice when it freezes.

12. <u>Icebergs come from glaciers that originate in the mountains</u>. If a glacier moves to the sea, it can break apart and float away as icebergs.

13. <u>Glaciers start on mountains as the result of snow that never completely melts in the summer</u>. If enough snow piles up, the weight causes it to slide down the mountain as a glacier.

14. <u>The captain is worried because 90% of the glacier is underwater and therefore not visible</u>. The captain steered clear of the visible part, but the underside of the boat could still hit the part that is underwater.

15. <u>The water table will be deeper in the area with lots of trees</u>. Since there are no trees to take away soil moisture in the one area, and since they each have the same kind of grass, then the area with the trees depletes soil moisture faster than the other one. As a result, there will be more unsaturated soil in the region with trees, and the water table will therefore be deeper.

16. <u>The air will cool as it expands</u>. That's what adiabatic cooling is all about.

17. Like the cloud in Experiment 5.3, <u>the fog will be thicker in the dusty area</u>.

18. Like the cloud in Experiment 5.3, <u>adiabatic</u> cooling accounts for most cloud formation.

19. <u>Groundwater pollution</u> is hard to track back because there is no accurate way to tell where the polluted groundwater came from.

SOLUTIONS TO THE MODULE #6 STUDY GUIDE

1. a. <u>Sediment</u> - A deposit of sand and mineral fragments, usually laid down by water

b. <u>Sedimentary rock</u> - Rock formed when heat, pressure, and chemical reactions cement sediments together

c. <u>Earth's crust</u> - Earth's outermost layer of rock

d. <u>Igneous rock</u> - Rock that forms from molten rock

e. <u>Plastic rock</u> - Rock that behaves like something between a liquid and a solid

f. <u>Earthquake</u> - A trembling or shaking of the earth as a result of rock masses suddenly moving along a fault

g. <u>Fault</u> - The boundary between a section of moving rock and a section of stationary rock

h. <u>Focus</u> - The point along a fault where an earthquake begins

i. <u>Epicenter</u> - The point on the surface of the earth directly above an earthquake's focus

2. The earth is divided into the <u>atmosphere, hydrosphere, lithosphere, mantle, and core.</u>

3. We can directly observe the <u>atmosphere, hydrosphere, and lithosphere.</u>

4. <u>The Moho separates the lithosphere from the mantle, and the Gutenberg discontinuity separates the mantle from the core.</u>

5. The lithosphere is composed of <u>soil, sediment, and crust.</u>

6. <u>Sedimentary rock is formed when sediments are solidified through heat, pressure, and chemical reactions. Igneous rock forms when molten rock solidifies.</u>

7. The extremes in temperature and pressure <u>make the rock behave sometimes like a liquid and sometimes like a solid.</u> That's why we call it plastic rock.

8. <u>Scientists observe seismic waves</u> which are usually generated by earthquakes. The behavior of these waves tells us a lot about the makeup and properties of the mantle and core.

9. <u>The inner core is solid because of pressure freezing.</u> Even though the inner core is hotter than the outer core, it remains solid because the pressure is so great that it forces iron atoms close enough together to be solid.

10. <u>The magnetic field is generated in the earth's core.</u>

11. <u>The magnetic field is caused by a large amount of electrical flow in the core.</u>

12. <u>The dynamo theory says that the motion of the core is due to the rotation of the earth and random currents in the liquid that makes up the core. The rapid decay theory states that the motion of the core started as a consequence of how the earth formed and is slowing down.</u>

13. <u>The rapid decay theory has been used to accurately predict the magnetic fields of other planets.</u> The dynamo theory fails miserably at this.

14. <u>The rapid decay theory requires a global catastrophe in order to be consistent with the data that indicate the magnetic field of the earth has reversed several times.</u>

15. <u>The fact that the rapid decay theory requires a catastrophe like Noah's Flood and the fact that the rapid decay theory indicates an earth 10,000 years old or younger tend to make many scientists shy away from it.</u> This is unfortunate, as there are good reasons to believe both of them!

16. <u>Without the magnetic field, cosmic rays from the sun would hit the earth.</u> These rays would kill all life on the planet.

17. <u>The plates are large "islands" of the earth's crust.</u> These plates float around on the plastic rock of the mantle.

18. <u>They can grind together to form earthquakes, or they can push up against one another to form mountains, or one plate can slide under another and form a trench.</u>

19. <u>Pangaea is a hypothetical supercontinent that might have existed in earth's past.</u> At one time, all of the continents might have fit together to form this supercontinent.

20. <u>Many good scientists ignore plate tectonics because it is typically linked to the idea of an earth that is billions of years old.</u> This is unfortunate because there is no reason to believe that the continents always moved slowly. Indeed, in a catastrophe like Noah's Flood, they could have moved very quickly.

21. <u>Earthquakes are caused by the motion of rock masses along a fault.</u>

22. <u>In the elastic rebound theory, a moving rock mass gets caught on the rough, jagged edge of a fault. The stationary rock on the other side of the fault starts to bend as the moving rock keeps trying to move. At some point, the stress becomes too great, and the moving rock breaks free, causing the rock on both sides of the fault to snap back into their original shape. The resulting vibrations are what we feel as an earthquake.</u>

23. For every one step up in the Richter scale, the power of the earthquake multiplies by 32. The first earthquake measured 4, and the second measured 8. The second earthquake was 4 units

higher, which means it released 32x32x32x32 = <u>1,048,576 times more energy than the first</u>!

24. <u>The four types of mountains are: volcanic mountains, domed mountains, fault-block mountains, and folded mountains. Volcanic and domed mountains need magma from the earth's mantle, fault-block mountains need vertical motion along a fault, and folded mountains need great pressure from two sides.</u>

SOLUTIONS TO THE MODULE #7 STUDY GUIDE

1. a. <u>Aphelion</u> - The point at which the earth is farthest from the sun

b. <u>Perihelion</u> - The point at which the earth is closest to the sun

c. <u>Coriolis effect</u> - The way in which the rotation of the earth bends the path of winds, sea currents, and objects that fly through different latitudes

d. <u>Air mass</u> - A large body of air with relatively uniform pressure, temperature, and humidity

e. <u>Weather front</u> - A boundary between two air masses

2. <u>The weather changes from day to day, while the climate does not</u>. Climate is what you generally expect from a region, while weather is what actually happens from day to day.

3. The three main factors are <u>thermal energy, uneven distribution of energy, and water vapor in the atmosphere</u>.

4. There is no answer for this one. Just be sure that given a picture or drawing of a cloud, you can determine which of the four major types of cloud it is.

5. When a cloud is dark, you add a suffix of "nimbus" or prefix of "nimbo." The proper term is <u>nimbostratus</u>, but stratonimbus is also correct.

6. When a cloud is higher than usual, you add the "alto" prefix. Thus, it would be an <u>altolenticular cloud</u>.

7. Insolation stands for <u>incoming solar radiation</u>.

8. At the summer solstice, the days in the Northern Hemisphere are at their longest, because the Northern Hemisphere is pointed towards the sun. They then begin to decrease so that at the autumnal equinox, they are exactly 12 hours long. Thus, <u>the days are greater than 12 hours long but are decreasing</u>.

9. In the Southern Hemisphere, the days get longer from the summer solstice to the winter solstice. At the autumnal equinox, they are exactly 12 hours long. Thus, <u>from the summer solstice to the autumnal equinox</u>, the days get longer but are still under 12 hours.

10. The Northern Hemisphere is pointed toward the sun at <u>aphelion</u>, so that's when it's summer in that hemisphere.

11. <u>If the earth had no axial tilt, the difference in temperature between night and day would be too severe for life to exist</u>.

12. <u>Temperature imbalances cause winds</u>.

13. There isn't a steady stream of wind blowing from the poles to the equator because <u>the temperature of the air changes as it changes latitude</u>. This causes loops of wind to develop at different latitudes.

14. <u>The Coriolis effect bends the wind patterns</u>.

15. Along the surface of the earth, winds blow from cold to warm. Thus, <u>the wind will blow from the mountain to the valley</u>.

16. Since it is continental, the <u>humidity is low</u>. Since it is polar, the <u>air mass is cold</u>.

17. Since it is maritime, the <u>humidity is high</u>. Since it is tropical, the <u>air mass is warm</u>.

18. This kind of weather is indicative of a <u>warm front</u>. It is not a stationary front because the rain would have lasted several days.

19. This kind of cloud progression is caused by an <u>occluded front</u>.

20. This kind of cloud pattern and resulting rain is indicative of a cold front. Thus, the <u>temperature should decrease after the rain</u>.

Weather Information Sources:

- Most local papers have a weather section that lists the data you are interested in.
- You can also watch (or tape) your local newscast on TV. They usually present the relevant data for yesterday.
- Libraries carry many more newspapers than what you get. You can look through the newspapers they have and find one that has most or all of the info you need. Check with the librarian to see if they keep the newspapers for a while, and if so, you need not go every day.

On the **INTERNET**, you can go to:

- **http://weather.noaa.gov/index.html** is a weather site run by the government. Select a state under "United States Weather" and then choose the nearest city under "Current weather conditions." That will lead you to a site which lists the current conditions and a 24-hour summary.
- **http://weather.unisys.com/surface/meteogram** is a website where you will find METEOGRAMS for many major cities. Choose a city near you and you will get a graph which plots the temperature, cloud cover, and pressure for the past 24 hours. Click on the "More Information" link to learn how to read meteograms.

SOLUTIONS TO THE MODULE #8 STUDY GUIDE

1. a. <u>Updraft</u> - A current of rising air

b. <u>Insulator</u> - A substance that does not conduct electricity very well

2. <u>The Bergeron process begins with cold clouds, while the collision-coalescence theory begins with warm clouds.</u>

3. <u>The Bergeron process describes precipitation from the top of cumulonimbus clouds, while the collision-coalescence theory describes precipitation from nimbostratus.</u> Remember, the top of a cumulonimbus cloud is near the top of the troposphere, where water freezes. Nimbostratus clouds, however, are much lower, so the water in those clouds is not as likely to be frozen.

4. <u>Only the size of the raindrop.</u> Drizzle has very small water droplets, while raindrops are bigger.

5. <u>Sleet is much smaller than hail, but both of them are frozen before they hit the ground. Freezing rain, on the other hand, is liquid until it hits a cold surface.</u> Hail and sleet also form differently, since hail is recycled through the cloud several times while sleet is not.

6. <u>The dew point is coldest on the second day.</u> It takes a colder temperature to form dew from air that is less humid or is lower in pressure.

7. <u>The first stage is the cumulus stage, where there is only an updraft and no precipitation. In the second stage, the mature stage, there are both updrafts and downdrafts as well as precipitation. The last stage, the dissipation stage, has only downdrafts and precipitation.</u>

8. <u>The thunderstorm is probably made up of several cells.</u> The mature stage of a typical thunderstorm cell lasts no longer than 30 minutes.

9. <u>The charge imbalance first forms in the cumulonimbus cloud, and it is due to water droplets or ice crystals rubbing against each other in glancing collisions.</u>

10. <u>The return stroke is responsible for the majority of light and sound in a lightning bolt.</u>

11. <u>Thunder is the result of superheated air traveling out from the lightning bolt in waves.</u> When those waves hit our eardrum, we interpret them as sound. Since the waves are violent, the sound is loud.

12. <u>Lighting strikes tall things because the positive charges in the ground tend to pile up in a tall object</u> because that's how they can get closest to the cloud.

13. <u>Sheet lightning is cloud-to-cloud lightning while lightning bolts are cloud-to-ground lightning</u>. The lightning bolts, therefore, hit the ground while sheet lightning never does.

14. <u>A cumulonimbus cloud must be present to form a tornado</u>. The vortex will not form without the strong updraft of a thunderstorm cell that forms a cumulonimbus cloud.

15. <u>The stages of a tornado are: the whirl stage, the organizing stage, the mature stage, the shrinking stage, and the decaying stage. The tornado is most destructive in its mature stage</u>.

16. <u>A hurricane starts out as a tropical disturbance, then becomes a tropical depression, then a tropical storm, and finally a tropical cyclone. The wind speeds in the storm determine in which classification a storm belongs</u>.

17. <u>Within the eye, a hurricane is calm</u>. It is often sunny as well.

18. <u>The Coriolis effect</u> causes hurricanes in different hemispheres to rotate differently.

19. <u>The atmospheric pressures are equivalent</u>. Even though they are far away from each other, they are on the same isobar, indicating equal pressure.

20. <u>The atmospheric pressure in Houston is lower</u>. Houston is 3 isobars from the "L" symbol, while Atlanta is 4 isobars away. This means Atlanta's pressure is higher.

21. The occluded front has triangles and ovals on the same side. That's nearest <u>San Francisco</u>.

22. The warm front has only ovals on it, and the ovals point in the direction of travel. Thus, <u>Indianapolis will get warmer weather soon</u>.

23. <u>Houston, TX</u> is near a cold front, so it might have thunderstorms right now.

24. <u>San Francisco, CA</u> is behind an occluded front. Since the weather described is that of an occluded front, San Francisco might have just experienced such weather.

SOLUTIONS TO THE MODULE #9 STUDY GUIDE

1. a. <u>Reference point</u> - A point against which position is measured

b. <u>Vector quantity</u> - A physical measurement that contains directional information

c. <u>Scalar quantity</u> - A physical measurement that does not contain directional information

d. <u>Acceleration</u> - The time rate of change of an object's velocity

e. <u>Free fall</u> - The state of an object that is falling towards the earth with nothing inhibiting its fall

2. In order for motion to occur, an object's position must change. Since this object's position is not changing, <u>it is not moving relative to the reference point</u>.

3. <u>The glass of water is moving relative to many reference points</u>. To someone standing still in the house, the glass is not moving. However, relative to any object not on earth, it is in motion. In fact, if you walk towards the glass, the glass is in motion relative to you, because the glass's position relative to you changes!

4. a. <u>The child is in motion relative to the two girls</u>. Even though the child is sitting still, his position relative to the girls is changing. Thus, he is in motion relative to the girls.

b. <u>The first girl is in motion relative to the child</u>. Since the position of her relative to the child is changing, she is in motion relative to him.

c. <u>The second girl is motionless relative to the first girl</u>. The girls are keeping perfect pace. Thus, their position relative to each other does not change. They are therefore both motionless with respect to each other.

5. This problem gives us distance and time and asks for speed. Thus, we need to use Equation (9.1). The problem wants the answer in miles per hour, however. We are given the time in minutes. Thus, we must make a conversion first:

$$\frac{30 \ \cancel{minutes}}{1} \times \frac{1 \ hour}{60 \ \cancel{minutes}} = 0.5 \ hours$$

Now we can use our speed equation:

$$speed = \frac{10 \ miles}{0.5 \ hours} = 20 \ \frac{miles}{hour}$$

6. This is another speed problem, but in this case, both of our units are wrong. We need meters per second, but we have kilometers and minute. Thus, we need to make two conversions:

$$\frac{6 \cancel{km}}{1} \times \frac{1000 \text{ meters}}{1 \cancel{km}} = 6000 \text{ meters}$$

$$\frac{45 \cancel{minutes}}{1} \times \frac{60 \text{ seconds}}{1 \cancel{minute}} = 2700 \text{ seconds}$$

Now we can use *those* numbers in our speed equation:

$$\text{speed} = \frac{6000 \text{ meters}}{2700 \text{ seconds}} = 2.2 \frac{\text{meters}}{\text{second}}$$

7. a. This is a <u>scalar</u> quantity since it has no direction. Meters is a distance unit, so this is <u>distance</u>.

b. This is a <u>vector</u> quantity, because it has direction in it. The units are distance over time squared, which is <u>acceleration</u>.

c. This is a <u>scalar</u> quantity. It has no direction. The units indicate it is <u>speed</u>.

d. This is a <u>scalar</u> quantity. It has no direction. Liters is a volume unit, so it is <u>none of these</u>.

e. This is a <u>vector</u> quantity, because it has direction in it. The units are distance over time, which is <u>velocity</u>. It is not speed because speed is not a vector quantity.

f. This is a <u>scalar</u> quantity. It has no direction. The units indicate it is <u>speed</u>.

8. As the picture shows, the car is behind the truck, but they are both traveling in the same direction. Thus, we get their relative velocity by subtracting their individual velocities:

relative velocity = 57 miles per hour - 45 miles per hour = 12 miles per hour

Since the car is traveling faster than the truck, it is catching up to the truck. Thus, the relative velocity is <u>12 miles per hour towards each other</u>.

9. Since the velocity is not changing, <u>the acceleration is zero</u>. The time was just put in there to fool you. Remember, acceleration is the change in velocity. With no change in velocity, there is no acceleration.

10. The initial velocity is 0, and the final velocity is 12 meters per second east. The time is 2 seconds. This is a straightforward application of Equation (9.2).

$$acceleration \ = \ \frac{final \ velocity \ - \ initial \ velocity}{time}$$

$$acceleration \ = \ \frac{12 \ \frac{meters}{second} - 0 \ \frac{meters}{second}}{2 seconds} \ = \ \frac{12 \ \frac{meters}{second}}{2 \ seconds} \ = \ 6 \ \frac{meters}{second^2}$$

Since the speed increased, the acceleration is in the direction of motion, 6 m/sec² east.

11. This is another application of Equation (9.2), because we are given time (12 minutes), initial velocity (30 miles per hour south) and final velocity (0, because it stops). We can't use the equation yet, however, because our time units do not agree. We'll fix that first:

$$\frac{12 \ \cancel{minutes}}{1} \times \frac{1 \ hour}{60 \ \cancel{minutes}} = 0.2 \ hours$$

Now that we have all time units in agreement, we can really use the acceleration equation:

$$acceleration \ = \ \frac{final \ velocity \ - \ initial \ velocity}{time}$$

$$acceleration \ = \ \frac{0 \ \frac{miles}{hour} - 30 \ \frac{miles}{hour}}{0.2 \ hours} \ = \ \frac{-30 \ \frac{miles}{hour}}{0.2 \ hours} \ = \ -150 \ \frac{miles}{hour^2}$$

The negative sign means the acceleration is in the opposite direction as the velocity. Thus, the acceleration is 150 miles/hour² north.

12. The physicist is technically right. For an object to be in free fall, there can be nothing obstructing the object's fall. Air resistance is an obstruction, and all objects experience air resistance.

13. Even though the physicist is technically correct, the effect of air resistance is so small on heavy objects that it can be ignored.

14. Neither will hit first. They both hit together. Remember, gravity accelerates all objects the same. Without air, there is not air resistance, so both objects are in true free fall. As a result, they will fall at exactly the same speed.

15. The rock is in free fall, so we can use Equation (9.3). Since the problem wants the answer in meters, we need to use 9.8 meters per second² as the acceleration.

$$\text{distance} = \frac{1}{2} \cdot (\text{acceleration}) \cdot (\text{time})^2$$

$$\text{distance} = \frac{1}{2} \cdot (9.8 \ \frac{\text{meters}}{\text{second}^2}) \cdot (4.1 \ \text{seconds})^2$$

$$\text{distance} = \frac{1}{2} \cdot (9.8 \ \frac{\text{meters}}{\text{second}^2}) \cdot (16.81 \ \text{second}^2) = \underline{82.4 \ \text{meters}}$$

16. The rock is in free fall, so we can use Equation (9.3). Since the problem wants the answer in feet, we need to use 32 feet per second^2 as the acceleration.

$$\text{distance} = \frac{1}{2} \cdot (\text{acceleration}) \cdot (\text{time})^2$$

$$\text{distance} = \frac{1}{2} \cdot (32 \ \frac{\text{feet}}{\text{second}^2}) \cdot (7 \ \text{seconds})^2$$

$$\text{distance} = \frac{1}{2} \cdot (32 \ \frac{\text{feet}}{\text{second}^2}) \cdot (49 \ \text{second}^2) = \underline{784 \ \text{feet}}$$

17. Since the object covers less ground in each time interval, it is traveling slower in each time interval. This clearly means that the object is slowing down. When an object is slowing, its acceleration is in the opposite direction of its velocity.

SOLUTIONS TO THE MODULE #10 STUDY GUIDE

1. a. <u>Friction</u> - A force resulting from the contact of two surfaces. This force opposes motion.

b. <u>Kinetic friction</u> - The friction that exists between surfaces when at least one of those surfaces is moving relative to the other

c. <u>Static friction</u> - The friction that exists between surfaces when neither surface is moving relative to the other

2. <u>Newton's First Law</u> - The velocity of an object will not change unless the object is acted on by an outside force.

<u>Newton's Second Law</u> - When an object is acted on by an outside force, the strength of that force is equal to the mass of the object times the resulting acceleration.

<u>Newton's Third Law</u> - For every action, there is an equal and opposite reaction.

3. Newton's First Law of Motion tells us that an object will not change velocity until acted on by an outside force. Often, this force is friction. In this problem, once the ball is thrown, no forces (not even friction) are operating on the ball. Thus, even in a year, its velocity will still be <u>3.0 meters per second to the west</u>.

4. <u>The beanbag will not fall next to the tree. Instead, it will fall north of the tree.</u> This is once again an application of Newton's First Law. While it is in the boy's hand, the beanbag has a velocity going north. When the boy drops the beanbag, it will still have a velocity going north. Thus, as it falls, it will travel north. When it lands, then, it will be north of the tree. In fact, ignoring air resistance, when it hits the ground, it will be right next to wherever the boy is at that instant, because it will be traveling north with the boy's velocity.

5. <u>The beanbag will land next to the tree.</u> In this case, the beanbag has no velocity. It is at rest with the boy standing next to the tree. When the running boy taps the beanbag lightly, it simply falls to the ground.

6. <u>The boxes will slam into the front seat.</u> The boxes have the same velocity as the car. When the car stops, they continue to move with the same velocity. This makes them move forward relative to the car, slamming them into the front seat.

7. Remember, friction is caused by bumps and grooves in a surface. When the road gets wet, the grooves in the road get filled with water. This makes it harder for the bumps on the tires to fit into them, which reduces friction. Thus, <u>the water fills in the grooves in the road, reducing the amount that they can grip the bumps on the tires of a car.</u>

8. <u>The static frictional force is always greater than the kinetic frictional force</u>. When the car is stopped, the man must overcome static friction to get the car moving. Once moving, the man needs only to overcome kinetic frictional force.

9. Since the object is moving with a constant velocity, that tells us acceleration is zero. Since the total force exerted on an object is equal to the object's mass times its acceleration (Newton's Second Law), then the total force on the object is zero as well. This means that the child exerts enough force to counteract kinetic friction, but no more. We must be talking about kinetic friction because the toy is already moving. Thus, the child exerts a force of <u>10 Newtons to the east</u>.

10. Ignoring friction, the only force involved is the force that the father exerts. Thus, the acceleration is due entirely to that force, and Equation (10.1) will give us the force's strength:

$$F = (mass) \cdot (acceleration)$$

$$F = (20\,kg) \cdot \left(2.0 \ \frac{m}{sec^2}\right) = 40 \ \frac{kg \cdot m}{sec^2}$$

Since a Newton is the same as a kilogram meter per second^2, the father is pushing with a force of <u>40 Newtons north</u>.

11. Since it takes more than 25 Newtons to get the object moving, <u>the static frictional force is 25 Newtons</u>. Once it is moving, however, it accelerates at 0.1 meters per second^2. This means the total force on the object is:

$$F = (mass) \cdot (acceleration)$$

$$F = (15\,kg) \cdot \left(0.10 \ \frac{m}{sec^2}\right) = 1.5 \ \frac{kg \cdot m}{sec^2}$$

This is the total force on the object, however. This force is the combination of the applied force (20 Newtons) and the kinetic frictional force. Since the kinetic frictional force opposes motion, it is opposite of the applied force. This means it subtracts from the applied force. Thus, in order for the total force to be 1.5 Newtons, <u>the kinetic frictional force is 18.5 Newtons against the motion</u>. You could also say "east" as the direction, because the motion is west, so in this case, east is against the motion.

12. The first part is easy. If the static frictional force is 500 Newtons, <u>the worker must apply more than 500 Newtons of force to get the box moving</u>. To accelerate the box once it is moving, the total force must be:

$$F = (mass) \cdot (acceleration)$$

$$F = (500 \, kg) \cdot \left(0.10 \, \frac{m}{sec^2} \right) = 50 \text{ Newtons}$$

If the total force is 50 Newtons, and the kinetic frictional force takes 220 Newtons to overcome, then the worker must apply 270 Newtons of force to the south in order to accelerate the box.

13. Static friction keeps objects from moving. If the gardener had to exert slightly more than 100 Newtons of force to get the rock moving, then the static frictional force is 100 Newtons. Once it got moving, the gardener keeps it moving at a constant velocity eastward. This tells us that the acceleration is zero, which means the total force on the rock is zero. Thus, the gardener applies enough force to overcome the kinetic frictional force, but no more. The kinetic frictional force, then, must be 45 Newtons.

14. The total force on the truck can be calculated from the mass and acceleration:

$$F = (mass) \cdot (acceleration)$$

$$F = (710 \, kg) \cdot \left(0.20 \, \frac{m}{sec^2} \right) = 142 \text{ Newtons}$$

This force comes from the two men and friction. The two men are pushing in the same direction, so their forces add. Thus, together they exert 376 Newtons of force. The total force is 142 Newtons, however, so the frictional force, which subtracts from the force of the two men, must be 234 Newtons west. "Against the motion" is another acceptable direction for the force.

15. The equal and opposite force is exerted by the doghouse on the child.

16. The player exerts a force on the ball because the ball's velocity changed. This means there was an acceleration, which means a force was exerted on the ball. The equal and opposite force is exerted by the ball on the player and is evidenced by the pain that the player feels when he catches the ball.

17. The wall exerts a force of 20 Newtons west, because it is equal and opposite of the man's force.

SOLUTIONS TO THE MODULE #11 STUDY GUIDE

1. The four fundamental forces are <u>the gravitational force, the weak force, the strong force, and the electromagnetic force. The electromagnetic force and the weak force are really different facets of the same force.</u> Thus, some say there are only 3 fundamental forces in Creation.

2. <u>The weakest force is the gravitational force. The strongest one is the strong force.</u>

3.
   1. <u>All objects with mass are attracted to one another by the gravitational force.</u>

   2. <u>The gravitational force between two masses is directly proportional to the mass of each object.</u>

   3. <u>The gravitational force between two masses is inversely proportional to the square of the distance between those two objects.</u>

4. When the 10 kg mass is replaced by a 20 kg mass, the mass doubled. This doubled the gravitational force. The 6 kg mass is also doubled to 12 kg. This once again doubles the gravitational force. Thus, the total change is 2x2 = 4. <u>The new gravitational force, then, is 4 times larger than the old one.</u>

5. The only difference is that the distance between the objects was increased by a factor of 4. The gravitational force decreases when the distance between the objects increases. It is decreased according to the square of that increase. Thus, the change is $4^2 = 16$. <u>The new gravitational force, then, is 16 times smaller than the old one.</u>

6. When the 1 kg mass is replaced by a 5 kg mass, the mass went up by a factor of five. This increases the gravitational force by a factor of five. The 2 kg mass is doubled to 4 kg. This doubles the gravitational force. The distance between the objects was decreased by a factor of 3. The gravitational force increases with decreasing the distance between the objects. The increase goes as the square of the change in the distance. Thus, the change is $3^2 = 9$. Therefore, the total change is 5x2x9 = 90. <u>The new gravitational force, then, is 90 times larger than the old one.</u>

7. <u>The equal and opposite force is the gravitational force that the moon exerts on the earth.</u>

8. <u>Centripetal force is required for circular motion.</u>

   <u>Centripetal force</u> - Force that is always directed perpendicular to the velocity of an object

9.

    1. <u>Circular motion requires centripetal force.</u>

    2. <u>The larger the centripetal force, the faster an object can travel in a circle.</u>

    3. <u>The larger the centripetal force, the *smaller* the circle of motion.</u>

10. The string is exerting the centripetal force. Thus, the force it exerts will change as needed by the principles of circular motion.

a. The string will need to exert <u>more force</u> because objects that travel fast need large centripetal force.

b. The string will exert <u>less force</u> because the larger the centripetal force, the smaller the circle. Thus, the larger circle needs less centripetal force.

11. <u>It is a myth. There is no such force.</u> The supposed effect of centrifugal force is just Newton's first law.

12. Traveling from "A" to "B" tells us that it is traveling clockwise. Its velocity is straight in the clockwise direction. Since it is traveling at a constant speed, the only force is centripetal, which always points to the center of the circle.

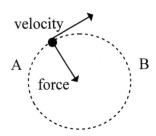

13. <u>Inner Planets</u> - Mercury, Venus, earth, and Mars

    <u>Outer Planets</u> - Jupiter, Saturn, Uranus, Neptune, Pluto

14. <u>Mercury, Venus, earth, Mars, Jupiter, Saturn, Uranus, Neptune, Pluto</u>

15. <u>Saturn, Uranus, Jupiter, and Neptune</u>

16. <u>Most of the asteroids are between the orbits of Mars and Jupiter.</u> This separates the inner planets from the outer planets.

17. <u>Perturbations in its orbit</u> cause an asteroid to become a meteor.

18. A comet is made of <u>a nucleus, a coma, and a tail. The nucleus is always present.</u>

19. All three parts of a comet are present <u>when the comet is close to the sun</u>.

20. <u>Comet orbits are elliptical</u>.

21. Physicists think that short-period comets come from a mass of comets called the <u>Kuiper belt</u>.

22. According to General Relativity, <u>gravity is caused by the fact that objects with mass bend the space they are in</u>.

23. According to the graviton theory, <u>gravity is caused by the exchange of particles called "gravitons."</u>

SOLUTIONS TO THE MODULE #12 STUDY GUIDE

1 a. <u>Photons</u> - Small "packages" of light that act just like small particles

b. <u>Charging by conduction</u> - Charging an object by allowing it to come into contact with an object which already has an electrical charge

c. <u>Charging by induction</u> - Charging an object by forcing some of the charges to leave the object

d. <u>Electrical current</u> - The amount of charge that travels through an electrical circuit each second

e. <u>Conventional current</u> - Current that flows from the positive side of the battery to the negative side. This is the way current is drawn in circuit diagrams, even though it is wrong.

f. <u>Resistance</u> - A measure of how much a metal impedes the flow of electrons

g. <u>Open circuit</u> - A circuit that does not have a complete connection between the two sides of the battery. As a result, current does not flow.

2. Like charges repel one another and will thus exert forces pushing the other directly away. Opposite charges attract and will therefore exert forces pulling the other directly in.

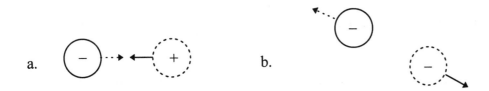

3. The electromagnetic force is inversely proportional to the distance between the objects. Thus, if the distance is increased by a factor of 3, <u>the force decreases by a factor of 9</u>. Since the poles are opposite, <u>it is an attractive force</u>.

4. The electromagnetic force is directly proportional to the charge. When the first charge is doubled, the force is doubled. Since the second charge is left the same, there is no change with respect to that charge. The force varies inversely with the square of the distance between objects. Thus, if the distance is halved, the force increases by a factor of four. The total change, then, is 2 x 4 = 8. <u>The new force is 8 times stronger than the old one</u>.

5. <u>The exchange of photons causes the electromagnetic force</u>.

6. <u>Charged particles do not glow because the photons they emit are not visible to you and me</u>. Under the right conditions, however, charged particles can emit visible light. At those times, you could say that the charged particles do "glow."

7. Charging by induction results in a charge opposite to that of the rod. Thus, <u>the object will be negatively-charged</u>.

8. Charging by conduction results in the same charge as the rod. Thus, <u>the object will be positively-charged</u>.

9. Voltage tells us what force the electricity source pushes with. The larger the voltage, the larger the force. This means the larger the voltage, the higher the energy of each electron. Thus, <u>each electron has high energy</u>. Current refers to how many electrons flow through the circuit. Thus, <u>few electrons flow through the circuit</u>. Even though there are few electrons, they each have high energy. Thus, <u>the circuit is dangerous</u>.

10. <u>A circuit is reasonably safe only when both the voltage and the current is low</u>. Please realize that "low" is a relative term. A low voltage is 9 volts or less. A low current is 0.001 Amps or less.

11. Conventional current flows from the positive side of a battery to the negative side.

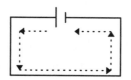

12. <u>Conventional current assumes that electricity is the flow of positive charges</u>. We know, however, that electricity is the flow of electrons, which are negative.

13. The longer the wire, the more chances the electrons have for colliding with atoms within the wire. Thus, <u>the longer wire has more resistance</u>.

14. The wider, or thicker, the wire, the more the electrons can spread out. This means there is less likelihood of electrons colliding with atoms in the wire. Thus, the thicker wire will have less resistance. Less resistance means less heat. Thus, <u>the thin wire will get hotter</u>. This is one of the leading causes of house fires. A person uses too thin an extension cord and tries to allow it to run too many devices. This draws too much current for the thin wire, heating it up to the point that it causes a fire.

15. In circuit (a), the open switch makes it impossible for any current to flow. Thus, the light bulb won't light. <u>In circuit (b), the light bulb lights</u> because the open switch is parallel to the light bulb. Thus, current can still flow through the bulb.

16. <u>The bulbs are wired in series</u> because a burnt-out bulb acts like an open switch. If the open switch turns off the bulbs, it is wired in series with the other bulbs.

17. <u>In a permanent magnet, the flow of charged particles is the motion of the electrons in its atoms</u>.

18. <u>As far as we know, magnets must always have both a north and south pole.</u>

19. <u>Yes,</u> it is possible. If the material responds strongly enough to a magnet, you can align its atoms and make it a magnet.

20. <u>If a material is not magnetic, its atoms cannot be aligned.</u> As a result, the flow of electrons is random, and the material cannot respond to a magnet.

SOLUTIONS TO THE MODULE #13 STUDY GUIDE

1. a. <u>Nucleus</u> - The center of an atom, containing the protons and neutrons

b. <u>Atomic number</u> - The number of protons in an atom

c. <u>Mass number</u> - The sum of the number of neutrons and protons in the nucleus of an atom

d. <u>Isotopes</u> - Two or more atoms that have the same number of protons but different numbers of neutrons

e. <u>Element</u> - A collection of atoms that all have the same number of protons

f. <u>Radioactive isotope</u> - An atom whose nucleus is not stable

2. The three constituents of the atom are the proton, neutron and electron. The electrons are significantly smaller than the other two, and the neutron is just slightly heavier than the proton. Thus, the order is <u>electron, proton, neutron</u>.

3. <u>The strong nuclear force</u> holds the protons and neutrons in the nucleus. <u>This force is caused by the exchange of pions between protons and/or neutrons.</u>

4. <u>The electromagnetic force</u> (or electroweak force) holds the electrons in orbit. They stay in orbit because they are attracted to the oppositely-charged protons.

5. <u>An atom is mostly empty space</u>.

6. The atomic number is defined as the number of protons in an atom. Atoms have the same number of electrons as they have protons. Thus, this atom has <u>34 electrons and 34 protons</u>. In order to get the symbol, we just look at the chart. The chart tells us that atoms with atomic number of 34 are symbolized with <u>Se</u>.

7. a. Since the chemical symbol is Ne, we can use the chart to learn that the atom has <u>10 protons</u>. This tells us there are also <u>10 electrons</u>. The mass number is the sum of protons and neutrons in the nucleus. Thus, there are also <u>10 neutrons</u>.

b. Since the chemical symbol is Fe, we can use the chart to learn that the atom has <u>26 protons</u>. This tells us there are also <u>26 electrons</u>. The mass number is the sum of protons and neutrons in the nucleus. Thus, there are 30 <u>neutrons</u>.

c. Since the chemical symbol is La, we can use the chart to learn that the atom has <u>57 protons</u>. This tells us there are also <u>57 electrons</u>. The mass number is the sum of protons and neutrons in the nucleus. Thus, there are <u>82 neutrons</u>.

d. Since the chemical symbol is Mg, we can use the chart to learn that the atom has <u>12 protons</u>. This tells us there are also <u>12 electrons</u>. The mass number is the sum of protons and neutrons in the nucleus. Thus, there are <u>12 neutrons</u>.

8. In order to be isotopes, the two atoms must have the same number of protons. Thus, <u>the second atom also has 18 protons</u>.

9. Isotopes must all have the same number of protons but different numbers of neutrons. Since the chemical symbol tells you how many protons that an atom has, in the end, only atoms with the same chemical symbol can be isotopes of one another. In order to be isotopes, then, the atoms must have the same chemical symbol but different mass numbers. Thus, <u>^{112}Sn, ^{124}Sn, and ^{120}Sn</u> are isotopes of one another.

10. All atoms symbolized with "O" have 8 protons according to the chart. This also means there are 8 electrons. Two of them can go into the first Bohr orbit, but we will have to put the remaining 6 in the second Bohr orbit. That's fine, because the second Bohr orbit can hold up to 8 electrons. The mass number indicates that there are also 8 neutrons:

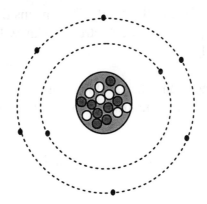

11. All atoms symbolized with "Mg" have 12 protons according to the chart. This also means there are 12 electrons. Two of them can go into the first Bohr orbit, and 8 can go in the second Bohr orbit. We will have to put the remaining 2 in the third Bohr orbit. That's fine, because the third Bohr orbit can hold up to 18 electrons. The mass number indicates that there are 13 neutrons:

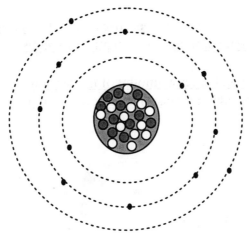

12. All uranium atoms, regardless of their mass number, have 92 protons and 92 electrons. That's what the periodic chart tells us for any element symbolized with "U." The first 2 electrons will go in the first Bohr orbit. The next 8 will go in the second Bohr orbit. The next 18 will go in the third Bohr orbit, and the next 32 will go in the fourth Bohr orbit. That makes 60 electrons. The remaining 32 will all fit in the fifth Bohr orbit because it can hold up to 50 electrons. Thus, the largest Bohr orbit is the fifth one, and there are 32 electrons in it.

13. The strong nuclear force is governed by the exchange of pions. Since pions have a very short lifetime, the strong nuclear force can only act over very tiny distances.

14. a. ^{98}Tc has 43 protons according to the chart. This means there must 55 neutrons. In beta decay, a neutron turns into a proton. This will result in an atom with 44 protons and 54 neutrons, or ^{98}Ru.

b. ^{125}I has 53 protons according to the chart. This means there must be 72 neutrons. In beta decay, a neutron turns into a proton. This will result in an atom with 54 protons and 71 neutrons, or ^{125}Xe.

15. a. ^{212}Bi has 83 protons according to the chart. This means there must be 129 neutrons. In alpha decay, the nucleus loses 2 protons and 2 neutrons. This will result in an atom with 81 protons and 127 neutrons, or ^{208}Tl.

b. ^{224}Ra has 88 protons according to the chart. This means there must be 136 neutrons. In alpha decay, the nucleus loses 2 protons and 2 neutrons. This will result in an atom with 86 protons and 134 neutrons, or ^{220}Rn.

16. Only gamma decay does not affect the number of neutrons and protons in a radioactive isotope.

17. In 1600 years, the 10 grams will be cut in half to 5 grams. In the next 1600 years, that 5 grams will be cut in half to 2.5 grams. That's a total of 3200 years, so the answer is 2.5 grams.

18. In one hour, the ^{11}C will have passed through three half-lives. During the first half-life, the 1 gram sample will be reduced to 0.5 grams. During the next half-life, that 0.5 grams will be reduced to 0.25 grams. In the final half-life, the 0.25 grams will be reduced to 0.125 grams.

19. Radioactive dating is usually unreliable because assumptions must be made as to the original condition of the object. These assumptions are usually erroneous.

20. Alpha particles pass through the least amount of matter before stopping, beta particles are next, and gamma rays pass through the most matter before stopping.

SOLUTIONS TO THE MODULE #14 STUDY GUIDE

1. a. <u>Transverse wave</u> - A wave whose propagation is perpendicular to its oscillation

b. <u>Longitudinal wave</u> - A wave whose propagation is parallel to its oscillation

c. <u>Supersonic speed</u> - Any speed that is faster than the speed of sound in the substance of interest

d. <u>Sonic boom</u> - The sound produced as a result of aircraft traveling at or above Mach 1

e. <u>Pitch</u> - The highness or lowness of a sound

2. <u>The engineers need to adjust the electronics to emit sound waves with shorter wavelengths.</u>
Remember, wavelength and frequency are inversely proportional. If the engineers want higher
pitch, they want larger frequencies, which means they want smaller wavelengths.

3. To determine the speed of sound, we use Equation (14.2):

$$v = (331.5 + 0.60 \cdot 30) \; \frac{m}{sec}$$

$$v = \underline{349.5 \; \frac{m}{sec}}$$

4. To determine a wave's frequency, we use Equation (14.1):

$$f = \frac{v}{\lambda}$$

$$f = \frac{349.5 \; \frac{m}{sec}}{0.5 \; m} = 699 \; \frac{1}{sec}$$

The frequency is <u>699 Hz</u>.

5. Infrasonic waves have frequencies less than 20 Hz (the lowest frequencies that human ears
can hear). Ultrasonic waves have frequencies of more than 20,000 Hz (the highest frequencies
that human ears can hear). Sonic waves have frequencies between 20 and 20,000 Hz. To
determine whether a wave is sonic, infrasonic, or ultrasonic, then, we must determine its
frequency. That's what we can do with Equation (14.1):

$$f = \frac{v}{\lambda} = \frac{345 \ \frac{m}{sec}}{500 \ m} = 0.69 \ \frac{1}{sec}$$

Since the frequency is less than 20 Hz, this is an <u>infrasonic wave</u>.

6. <u>The physicist will not be able to hear the alarm because, without air, the sound waves from the alarm have nothing through which to travel. Thus, they cannot make waves. As a result, there is no sound.</u>

7. <u>Sound waves are longitudinal waves.</u>

8. To determine the distance, we will use the time difference between the lightning flash and the sound. We will assume that the light from the lightning reaches our eyes essentially at the same time as the lightning was formed. Thus, the time it takes for the sound to travel to you will determine the distance. First, then, we need to know the speed of sound:

$$v = (331.5 + 0.60 \cdot 13) \ \frac{m}{sec}$$

$$v = 339.3 \ \frac{m}{sec}$$

Now we can use Equation (14.3):

$$distance = (speed) \times (time)$$

$$distance = (339.3 \ \frac{m}{sec}) \times (2.3 \ sec) = \underline{780.39 \ m}$$

9. Sound travels faster in solids than it does in gases or liquids. Thus, <u>the sound travels faster in the wall</u>.

10. Remember, when a wave strikes an obstacle, part of the wave is reflected and part of it is transmitted through the obstacle. Thus, only a portion of the wave actually starts traveling through the wall. This means <u>the amplitude of the wave will be smaller</u> because only part of the wave is present in the wall.

11. To determine the speed of the jet, we first have to determine the speed of sound. After all, Mach 2.5 means 2.5 times the speed of sound. Thus, we need to know the speed of sound in order to determine the speed of the jet.

$$v = (331.5 + 0.60 \cdot T) \; \frac{m}{sec}$$

$$v = (331.5 + 0.60 \cdot 1) \frac{m}{sec} = 332.1 \; \frac{m}{sec}$$

Since sound travels at 332.1 m/sec, Mach 2.5 is 2.5 x (332.1 m/sec) = <u>830.25 m/sec</u>.

12. This is much like the previous problem. To determine the Mach, we need to first determine how quickly sound travels in that air:

$$v = (331.5 + 0.60 \cdot T) \; \frac{m}{sec}$$

$$v = (331.5 + 0.60 \cdot 0) \frac{m}{sec} = 331.5 \; \frac{m}{sec}$$

To determine the Mach, then, we can just divide the speed of the jet by the speed of sound. This will tell us how many times faster than sound the jet is traveling:

$$464.1 \div 331.5 = 1.4$$

This means that the jet is traveling at <u>Mach 1.4</u>.

13. <u>When a jet travels at Mach 1 or higher, it emits a shock wave of air that causes a very loud boom. This boom can damage people's ears and buildings.</u> Thus, sonic booms must be avoided when people or buildings are nearby.

14. Since the string becomes shorter, the wavelength will be smaller. This will result in a larger frequency. Thus, <u>the pitch will be higher</u>.

15. <u>The wavelength, frequency, and speed will all be the same</u>. After all, the pitch is determined by the frequency, which, in turn, determines the wavelength. The speed depends only on the temperature. <u>The amplitudes of the waves will be different</u>, however, because amplitude determines loudness.

16. As the car travels away from you, the sound waves that are produced by the horn get farther and farther apart. This makes the wavelength seem longer to your ears, which will result in a smaller frequency. Thus, <u>the horn's pitch will get lower</u>.

17. As you travel towards the police car, you will encounter the crests of the sound waves faster than when you stand still. As a result, the wavelength will seem shorter to you, which will produce larger frequencies. This means that the pitch will get <u>higher</u>.

18. The bel scale states that every bel unit corresponds to a factor of ten in the intensity of the sound waves. Thus, we need to determine how many bel units the sound of the traffic is, as compared to the sound of your voice:

$$\frac{80 \text{ decibels}}{1} \times \frac{1 \text{ bel}}{10 \text{ decibels}} = 8 \text{ bels}$$

$$\frac{100 \text{ decibels}}{1} \times \frac{1 \text{ bel}}{10 \text{ decibels}} = 10 \text{ bels}$$

Since the traffic is 2 bels louder than normal conversation, the increase in sound wave intensity is 2 factors of ten higher. Thus, the traffic has sound waves with intensities that are 10 x 10 = 100 times larger than the intensities of sound waves from your voice.

19. The bel scale states that every bel unit corresponds to a factor of ten in the intensity of the sound waves. Thus, we need to determine how many bel units were fed into the amplifier:

$$\frac{30 \text{ decibels}}{1} \times \frac{1 \text{ bel}}{10 \text{ decibels}} = 3 \text{ bels}$$

If the amplifier magnifies the intensities of the waves by a factor of 1,000, that's the same as 10 x 10 x 10. Thus, the sound coming out of the amplifier will be 3 bels larger, as each factor of ten represents one increase in the bel level. Thus, the sound coming out will be 6 bels, which is the same as 60 decibels.

SOLUTIONS TO THE MODULE #15 STUDY GUIDE

1. a. <u>Electromagnetic wave</u> - A transverse wave composed of an oscillating electrical field and a magnetic field that oscillates perpendicular to the electrical field

b. <u>The Law of Reflection</u> - The angle of reflection equals the angle of incidence

2. <u>The wave theory of light views light as two transverse waves, one made of an oscillating magnetic field and the other an oscillating electrical field. The particle theory of light views a ray of light as a beam of individual particles called photons. The quantum-mechanical theory says that light is both a particle and a wave. It is made up of individual packets that behave like particles, but the packet is actually made up of a wave.</u>

3. <u>Light waves oscillate a magnetic field and an electrical field.</u> Each one oscillates perpendicular to the other, as well as perpendicular to the direction of travel.

4. Einstein's special theory of relativity says that <u>nothing can travel faster than the speed of light in any given substance.</u>

5. Unlike sound, light travels slower in liquids than gases. Thus, <u>the light's speed increased once it left the water.</u>

6. The acronym ROY G. BIV allows us to remember the relative size of the colors' wavelengths. Red is longest and violet is shortest. Thus, in terms of increasing wavelength, the colors are: <u>violet, green, yellow, and orange.</u>

7. Wavelength and frequency are inversely proportional. Thus, in terms of increasing frequency, it is: <u>orange, yellow, green, and violet.</u>

8. Radio waves have wavelengths longer than visible light while X-rays have shorter wavelengths. This comes from Figure 15.3. Just like the visible light colors, you need not memorize any values for the wavelengths, but of the major categories in the figure, you need to know their relative wavelengths. Since frequency and wavelength are inversely proportional, <u>radio waves have lower frequencies than visible light while X-rays have higher frequencies.</u>

9. <u>Infrared light is not visible.</u> Thus, even though human bodies constantly emit infrared light, we cannot see that happening. There are special devices you can get that do, indeed, detect the infrared light that the human body emits. This allows you to see living organisms and other hot objects, even in the darkest of nights.

10. By the Law of Reflection, <u>the reflected light also makes a 15 degree angle relative to that line</u>.

11. <u>Yes</u>. Remember, to see a part of his body, light must be able to travel from that part of his body, reflect off the mirror, and hit his eye. His brain will then extend that line backwards,

forming an image in the mirror. The only constraint is that the light which strikes the mirror must obey the Law of Reflection:

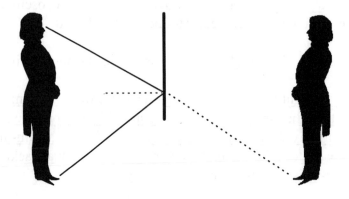

Had the mirror been much smaller, there would have been no way light could travel from his foot and hit his eyes and also obey the Law of Reflection:

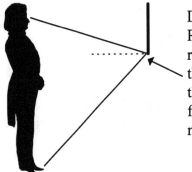

Does NOT obey the Law of Reflection. The angle of reflection is clearly smaller than the angle of incidence. Thus, there is no way light can travel from his foot, reflect off the mirror, and hit his eye!

12. <u>The light ray can be reflected or refracted.</u>

13. When light travels from a substance in which it moves quickly to a substance in which it moves slowly, the light bends towards the perpendicular. Since light moves faster in air than in glass, <u>the light will bend towards the perpendicular line.</u>

14. In order for you to see the objects underwater, light must travel from the object to your eyes. Thus, the light must travel out of the water and into the air. <u>When light travels from water to air, it bends. This causes your mind to form a false image of the object in a slightly different location.</u>

15. <u>There must be water droplets suspended in the air, the sun must be shining on them from behind you, and the sun must be at a certain angle (or height in the sky).</u> The water droplets cause the refraction. In order to separate the colors enough to see them, however, the light must be refracted, reflected, and refracted again (see Figure 15.6). To do that, light must enter the water droplet on the side from which you are viewing it.

16. A converging lens causes light rays to bend so that they converge to a single point. Diverging lenses cause light rays to bend away from each other.

17. The function of a lens depends on its shape. Lens (a) is a converging lens and lens (b) is a diverging lens.

18. The eye focuses light by changing the shape of its lens. A camera focuses light by moving the position of the lens. The eye's method is faster and much more precise.

19. If your red cone cells no longer worked, your brain would think that you never saw red light. When the white light reflected off of the paper and hit your eyes, then, your green cone cells would send signals to your brain, as would your blue cone cells. As a result, the white paper would appear to be blue-green or cyan. If you looked at a red piece of paper, it is red because it only reflects red light. Thus, it would send only red light to your eyes. You cannot see red light, however, so the red paper would appear to be black.

20. In order to look violet, it must absorb all colors except violet. Thus, it absorbs red, yellow, orange, green, blue, and indigo light.

21. Since cyan absorbs all colors except blue and green, it will absorb any red light shone on it. As a result, no light will make it to your eyes. In red light, then, the paper would look black. When you shined green light on it, the green light would be reflected. There would be no blue to mix with it, though. In green light, the paper would look green.

SOLUTIONS TO THE MODULE #16 STUDY GUIDE

1. a. <u>Nuclear fusion</u> - The process by which two or more small nuclei fuse to make a bigger nucleus

 b. <u>Nuclear fission</u> - The process by which a large nucleus is split into two smaller nuclei

c. <u>Critical mass</u> - The amount of isotope necessary to cause a chain reaction

d. <u>Star magnitude</u> - The brightness of a star on a scale of -8 to +17. The *smaller* the number, the *brighter* the star.

e. <u>Light year</u> - The distance light travels in one year

f. <u>Galaxy</u> - A massive ensemble of hundreds of millions of stars, all interacting through the gravitational force, orbiting around a common center

2. Starting on the inside, the sun is divided into <u>the core, the radiation zone, the convection zone, and the photosphere.</u>

3. <u>The sun gets its power from nuclear fusion that occurs in the core.</u>

4. We see <u>the photosphere.</u>

5. Since a large nucleus split into two smaller nuclei (and a few neutrons), this is <u>nuclear fission</u>.

6. Since two small nuclei became a bigger nucleus (plus a neutron), this is <u>nuclear fusion</u>.

7. In both processes, mass is converted into energy. Thus, <u>the mass of the starting materials is larger than the mass of the materials the process makes.</u>

8. It is impossible for a nuclear power plant to experience a nuclear explosion because <u>a power plant does not have a critical mass of ^{235}U or ^{239}Pu.</u>

9. <u>Nuclear fusion is a better means of producing energy because there are no radioactive by-products, there is no chance of meltdown, and the starting materials are cheap.</u>

10. <u>We cannot use nuclear fusion yet because we cannot master the technology that can make it economically feasible.</u>

11. To classify a star, you find where its magnitude and spectral letter put it on the H-R diagram. Taking the values given for the stars in this problem, you can come up with the following H-R diagram:

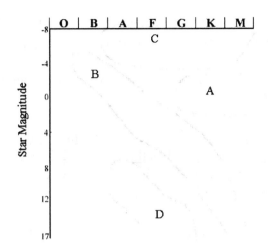

Thus:

a. Red giant   b. Main sequence   c. Supergiant   d. White dwarf

12. Our sun is a main sequence star. Thus, star (b) is most like our sun.

13. In general, supergiants are the largest, red giants are next, main sequence stars are next, and white dwarfs are the smallest. Thus, in terms of *increasing* size, it is (d), (b), (a), and (c).

14. Brightness is given by magnitude. The *smaller* the magnitude, however, the brighter the star. Thus, in terms of *increasing* brightness, it is (d), (a), (b), and (c).

15. The farther to the right on the H-R diagram, the cooler the star. Thus, (a) is the coolest.

16. All three of these are variable star types. Thus, their brightness changes radically with time.

17. The big difference between these star types is lifetime. Pulsating stars last a long time, supernovas exist very briefly, and novas are somewhere in between.

18. A nebula is a cloud of bright gases that are the remains of a supernova.

19. The two methods are the parallax method and the apparent magnitude method. The parallax method is exact, but the apparent magnitude method can be used to measure long distances.

20. Cepheid variables are important to measuring long distances because they seem to have a relationship between their period and their magnitude. That allows them to be used in the apparent magnitude method for measuring long distances in the universe.

21. The four galaxy types are spiral, lenticular, elliptical, and irregular. The Milky Way is a spiral galaxy.

22. Stars group together to form <u>galaxies</u>, which group together to form <u>groups</u>, which group together to form <u>clusters</u>, some of which group together to form <u>superclusters</u>.

23. The earth's solar system belongs in the <u>Milky Way</u>, which belongs to the <u>Local Group</u>, which belongs to the <u>Virgo Cluster</u>.

24. <u>Most astronomers believe that the universe is expanding because the light from nearly every galaxy experiences a red shift before it reaches the earth, and the red shift increases the farther the galaxy is from the earth</u>. This is evidence that the galaxies are all moving away from us, which would indicate an expanding universe.

25. <u>The universe could expand without a center, or it could expand with a center that is earth's solar system</u>.

26. <u>Yes, the way that the universe is expanding makes a great deal of difference. The theories that can be developed for the formation of the universe depend on that initial assumption</u>.

Tests

TEST FOR MODULE #1

1. Write out the definitions for the following terms:

a. Atom
b. Molecule
c. Concentration

2. Sulfur is a yellow powder that is composed of sulfur atoms. Sulfur dioxide is a colorless, poisonous gas that contains sulfur atoms. Is sulfur dioxide composed of atoms or molecules?

3. While looking at historical grave markers, you find a statuette that is blue-green in color. In order to read the inscription, you scrub the surface of the statuette, and the blue-green color comes off as a fine powder. What color is the statue underneath?

4. Which picture represents a bunch of atoms? Which represents a bunch of molecules?

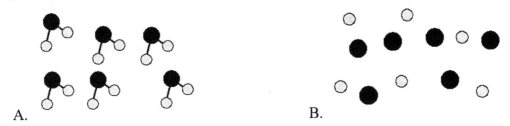

A.   B.

5. You are reading a scientist's notes and you notice a measurement that is listed as "12.3 kilograms." Does this measurement represent length, mass, or volume?

6. What metric prefix means "1,000"?

7. How many centimeters are in 1.6 meters?

8. An object's volume is 0.12 kL. What is its volume in liters?

9. A rock has a mass of 45.1 kg. What is its mass in slugs? (1 slug = 14.59 kg)

10. Ammonia is the active ingredient in many household cleaners. Suppose I were to make up two buckets of cleaner. In the first, I will take 5 cups of ammonia and add to that 45 cups of water. In the second, I will take 5 cups of ammonia and add 30 cups of water. Which bucket contains the most powerful cleaner?

TEST FOR MODULE #2

1. Define the following terms:

    a. Humidity
    b. Absolute humidity
    c. Relative humidity
    d. Greenhouse effect
    e. Parts per million

2. When does water evaporate more slowly, under conditions of high humidity or low humidity?

3. Will sweating help cool you down when the humidity is 100%?

4. What gas makes up the majority of the air we inhale?

5. What gas makes up the majority of the air we exhale?

6. A chemist is monitoring the rate at which a certain substance burns. The chemist burns the substance in a fireplace that uses the room's air supply. The chemist then repeats the experiment, this time in a chamber whose air mixture is 50% oxygen and 50% nitrogen. In which trial will the substance burn the fastest?

7. Why is it important to have ozone in earth's air?

8. For good health, should we increase or decrease the concentration of ground-level ozone in the air?

9. Is global warming happening today?

10. Convert 1% into ppm.

11. The concentration of nitrogen oxides in the air today is about 0.019 ppm. What is that in percent?

12. What pollutant concentration was decreased by the mandate of catalytic converters?

TEST FOR MODULE #3

1. Define the following terms:

a. Atmosphere
b. Barometer
c. Homosphere
d. Heterosphere
e. Jet streams
f. Heat
g. Temperature

Choose your answers to problems 2, 3, 5, 6, and 8 from the regions of the atmosphere listed below:

Regions of the atmopshere: exosphere, mesosphere, stratosphere, thermosphere, troposphere

2. If you want to study weather, which region of the atmosphere would you study?

3. If you want to study the ozone layer, which region of the atmosphere would you study?

4. If a sample of air is predominately oxygen, did it most likely come from the homosphere or the heterosphere?

5. Which regions of the atmosphere are in the homosphere?

6. Which regions of the atmosphere are in the heterosphere?

7. A barometer develops a leak in the column which is supposed to be free of air. As air seeps into the column, what will happen to the height of the liquid in that column?

8. In what region(s) of the homosphere does temperature increase with increasing altitude?

9. Why is the "ozone hole" a seasonal phenomenon that exists only at the South Pole?

10. We all know that ice melts because of heat. Why is it correct to say that ice also freezes because of heat?

11. If you were able to measure the speed of the molecules in the air while you were traveling up through the troposphere, would the speed of the molecules increase, decrease, or stay the same as your altitude increased?

TEST FOR MODULE # 4

1. Define the following terms:

a. Electrolysis
b. Polar molecule
c. Solvent
d. Solute
e. Cohesion
f. Hard water

2. What is the chemical formula for water?

3. Some metals tend to absorb oxygen but not hydrogen. Suppose such a metal was covering the battery in an electrolysis experiment like Experiment 4.1. Which is the more likely erroneous result: HO_2 or H_4O?

4. Why is water a liquid at room temperature when all other chemically similar substances are gases at room temperature?

5. Carbon dioxide (CO_2) is one of the gases that we exhale when we breathe. Carbon monoxide (CO) is a poisonous gas associated with burning things under conditions of low oxygen. How many atoms are in one molecule of CO_2? How many atoms are in one molecule of CO?

6. The principal component of gasoline is octane, whose molecules are composed of eight carbon atoms (C) and eighteen hydrogen atoms (H). What is the chemical formula?

7. Why are water molecules polar?

8. If a substance does not dissolve in water, is it ionic, polar, or nonpolar?

9. If a substance dissolves in water, will it dissolve in vegetable oil, a nonpolar substance?

10. Is hard water the result of a city's water treatment process?

TEST FOR MODULE #5

1. Define the following terms:

a. Transpiration
b. Condensation
c. Residence time
d. Percolation
e. Adiabatic cooling

2. Where does the majority of earth's water reside?

3. What is the largest source of liquid freshwater on the planet?

4. What water source is a molecule of water in once it has gone through transpiration?

5. Water was in the ocean and is now in a cloud. What two hydrologic cycle processes happened in order to make that transfer?

6. Where is the residence time longer: in the ocean or in a fast-moving stream?

7. If a lake has no means of getting rid of water except evaporation, does it contain saltwater or freshwater?

8. What do the oceans tell us about the age of the earth?

9. An enormous amount of ocean water in the polar region freezes. Does it form an iceberg? Why or why not?

10. What process in the hydrologic cycle is responsible for making glaciers?

11. What causes the condensation that makes most clouds?

12. If a sample of gas is compressed and nothing else is allowed to change, what will happen to the temperature of the gas?

13. If there is a lot more rain than normal in an area over an extensive length of time, what happens to the depth of the water table?

14. Why is groundwater pollution so hard to trace back to its original source?

TEST FOR MODULE #6

1. Define the following terms:

a. Sedimentary rock
b. Plastic rock
c. Fault
d. Epicenter

2. Label the sections (a-d) of the earth shown in the figure:

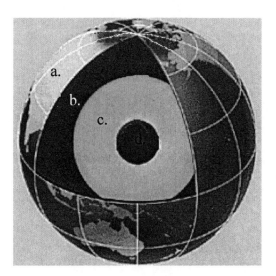

3. What have scientists observed in order to learn about earth's interior?

4. In the figure from problem #2, what two letters is the Moho between?

5. What causes the earth's magnetic field?

6. What two theories attempt to explain the earth's magnetic field? Which theory is the most scientifically valid?

7. What benefit do we derive from the earth's magnetic field?

8. In a survey of the deep ocean, sonar measurements detect a deep trench on the bottom that runs as far as the instruments detect. What is the most likely cause of the trench?

9. The earthquake activity of two regions on earth is measured. The first region sits near the middle of one of the plates in the earth's crust, while the other is very near a boundary between two plates. Which will (most likely) have the greatest earthquake activity?

10. Many powerful earthquakes are followed later by less-powerful earthquakes called "aftershocks." If an earthquake measures 6 on the Richter scale and is followed by an aftershock

that measures 4, how many times more energy was released in the original earthquake as compared to the aftershock?

11. If a region of the earth has a lot of volcanic activity, what kinds of mountains do you expect to find there?

12. If a region of the earth is on the boundary between two plates, what kind of mountains do you expect to find there?

TEST FOR MODULE #7

1. Define the following terms:

    a. Aphelion
    b. Perihelion
    c. Coriolis effect
    d. Air mass
    e. Weather front

2. Identify the clouds in the following pictures:

a.

b.

c.

3. Of the 3 main factors that influence weather, which is mostly responsible for winds?

4. What are dark cumulus clouds called?

5. If an area receives a large amount of insolation, is it likely to be warm or cold?

6. In the Southern Hemisphere, are the day lengths greater than or less than 12 hours between the winter solstice and the spring equinox? Are the day lengths increasing or decreasing during that time?

7. Why is the Northern Hemisphere in winter when the earth is closest to the sun?

8. What would happen to life on the planet if the earth had no axial tilt?

9. Without two specific factors, the global wind patterns would be simple. They would blow from the poles to the equator. What two factors shape the global winds into the complex patterns that we actually see?

10. What causes the wind in a certain region to be different than what we expect based on the global patterns we see?

11. An air mass is dry and warm. What kind of air mass is it?

12. You watch the sky as cirrus clouds form followed by stratus and nimbostratus clouds. Do you expect a violent rainstorm or a long, lighter rain?

TEST FOR MODULE #8

1. Define the following terms:

a. Updraft
b. Insulator

2. The same cloud precipitates snow on a mountain and rain in the nearby valley. Does the Bergeron process or the collision-coalescence theory best describe the process causing precipitation from that cloud?

3. What is the dew point? What two factors influence it?

4. A thunderstorm cell is raining, and there is no updraft. In what stage is the thunderstorm cell? Will there be hail at this point in the thunderstorm?

5. If the mature stage of a thunderstorm lasts for 30 minutes maximum, why can thunderstorms rain heavy sheets of rain for several hours?

6. Why don't you see lightning from nimbostratus clouds?

7. What happens first in a lightning bolt: a return stroke or a stepped leader?

8. How does lightning cause thunder?

9. Is it possible for sheet lightning to strike a person?

10. A tornado is in its organization stage. Has it touched the ground yet?

11. What differentiates a tropical storm from a tropical disturbance?

12. Where is the safest place in a hurricane?

(TEST CONTINUES ON THE NEXT PAGE)

Given the following weather map, answer questions 13 - 16.

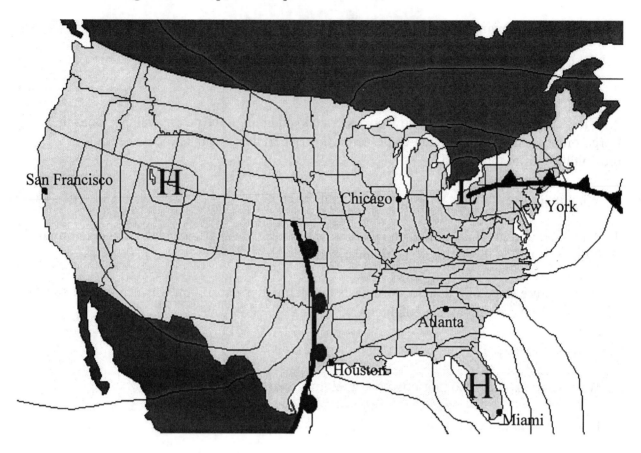

13. Is the atmospheric pressure in Houston, TX higher, lower, or equivalent to that in Atlanta, GA?

14. Is the atmospheric pressure in Chicago, IL higher, lower, or equivalent to that in New York, NY?

15. What city listed on the map might have been experiencing thundershowers at the time this map was drawn?

16. What city listed on the map should expect warmer weather?

TEST FOR MODULE #9
(There are 60 seconds in a minute and 3600 seconds in an hour.)

1. Define the following terms:

    a. Reference point
    b. Vector quantity
    c. Scalar quantity
    d. Acceleration
    e. Free fall

2. Why must one use a reference point to determine whether or not an object is in motion?

3. After a visit to your grandmother's house, you get in your car to go home. You are in the front passenger's seat and your mother is driving. As you back out of your grandmother's driveway, she stands outside, waving good-bye.

    a. Who is in motion relative to you?
    b. Who is motionless relative to you?

4. How many miles per hour does a car travel if it makes a 40 mile trip in 30 minutes?

5. What is the velocity of a bicycle (in meters per second) if it travels 1 kilometer west in 4.1 minutes?

6. You are looking in a scientist's lab notebook and find the following unlabeled measurements. In each case, determine what physical quantity the scientist was measuring.

    a. 12.1 meters per second
    b. 31.2 feet
    c. 14 millimeters per hour to the left
    d. 4.5 yards per minute^2 north

7. An eagle swoops down to catch a baby rabbit. Luckily for the rabbit, he sees the eagle and runs. An all-out chase ensues with the rabbit running east at 5.4 meters per second and the eagle pursuing at 4.4 meters per second. What is the relative velocity of predator and prey?

4.4 meters/sec east

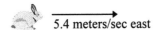
5.4 meters/sec east

8. What is the acceleration of an object that moves with a constant velocity?

9. A skier reaches the bottom of a slope with a velocity of 12 meters per second north. If the skier comes to a complete stop in 3 seconds, what was her acceleration?

10. A car goes from 0 to 60 miles per hour north in 5 seconds. What is the car's acceleration?

11. A person standing on a bridge over a river holds a rock and a ball in each hand. He throws the ball down towards the river as hard as he can and at the same time simply drops the rock. After both have left the person's hand, does one have a greater acceleration? If so, which one?

12. Why does a dropped feather hit the ground later than a rock dropped at the same time?

13. A physics student climbs a tree. To measure how high she has climbed, she drops a rock and times its fall. It takes 1.3 seconds for the rock to hit the ground. How many feet has she climbed?

TEST FOR MODULE #10

1. Define the following terms:

a. Friction
b. Kinetic friction
c. Static friction

2. State Newton's three laws of motion.

3. A pilot is flying a mission to drop bombs on an enemy airfield. The plane is flying high and fast to the north, and the city is due north. Should the pilot drop the bombs before the plane is over the airfield, when the plane is over the airfield, or after the plane has passed the airfield?

4. A cruel boy has placed a mouse on the outer edge of a disk. He slowly starts to spin the disk, accelerating it faster and faster until the disk and mouse are both spinning around at an alarming rate. What will happen to the mouse if the boy suddenly stops the disk without touching the mouse: will the mouse continue to spin like it was before; will the mouse stop with the disk; or will the mouse start moving straight, skidding off the disk?

5. An ice cube (mass = 1.0 kg) slides down an inclined serving tray with an acceleration of 4.0 meters per second^2. Ignoring friction, how much force is pulling the ice cube down the serving tray?

6. A baseball player (mass = 75 kilograms) is running north towards a base. In order to avoid being tagged by the ball, the baseball player slides into the base. If his acceleration in the slide is 5.0 meters per second^2 south, what is the kinetic frictional force between the baseball player and the ground?

7. A man pushes a heavy cart. If the man exerts a force of 200 Newtons on the cart to keep it moving at a constant velocity, what is the frictional force between the cart and the ground? Is this kinetic friction or static friction?

8. You are looking through a physicist's laboratory notebook and notice two numbers for the friction between a block of wood and a laboratory bench. The numbers are 8 Newtons and 11 Newtons. Which refers to static friction and which refers to kinetic friction?

9. A woman pulls on a stubborn dog (mass = 30 kilograms). The dog resists the pull with a force of 30 Newtons. In addition, the static friction between the dog and the ground is 20 Newtons, while the kinetic friction is 7 Newtons. How much force must the woman exert to get the dog moving? If the woman ends up dragging the dog with an acceleration of 1.0 meters per second^2 to the west, what force is she pulling with?

10. When baseball players hit the ball hard enough, their bats can sometimes break. What is exerting a force on the bat, causing it to break?

TEST FOR MODULE #11

1. A child is twirling a toy airplane on a string at a constant speed:

Draw the velocity of the plane and the force that it experiences.

2. Which is the weakest of the fundamental forces?

3. A student drops a ball, and it begins to fall due to the force of gravity that the earth exerts on it. What is the equal and opposite force demanded by Newton's Third Law of Motion?

4. The gravitational force between two objects (mass$_1$ = 5 kg, mass$_2$ = 2 kg) is measured when the objects are 10 centimeters apart. If the 5 kg mass is replaced with a 20 kg mass and the 2 kg mass is replaced with a 12 kg mass, how does the new gravitational attraction compare to the first one that was measured?

5. The gravitational force between two objects (mass$_1$ = 10 kg, mass$_2$ = 6 kg) is measured when the objects are 12 centimeters apart. If the distance between them is increased to 36 centimeters, how does the new gravitational attraction compare to the first one that was measured?

6. Two moons orbit different planets, but they orbit their planets at the same distance. If the first one takes 3 months to make an orbit and the second takes 1 year, which is being subjected to the greatest gravitational attraction?

7. List the outer planets.

8. Which planet in the solar system receives the least amount of insolation?

9. Which of the following orbits is more likely that of a comet?

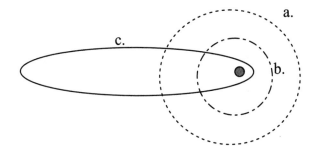

10. What part or parts of a comet come and go during a comet's orbit?

11. What is the Kuiper belt a source of?

12. What theory says that gravity is caused by the exchange of particles between objects with mass?

13. Suppose a scientist determines that there are only two fundamental forces in nature: the electroweak force and the strong force. Which of the two current theories of gravity does this mean is true?

TEST FOR MODULE #12

1. Define the following terms:

a. Photons
b. Charging by conduction
c. Charging by induction
d. Electrical current
e. Conventional current
f. Resistance
g. Open circuit

2. For the following situation, draw the force exerted by the solid object with a solid arrow. Draw the force exerted by the dashed object with a dashed arrow:

3. The force between the south pole of one magnet and the south pole of another magnet is measured. If the distance between those magnets is suddenly doubled, how will the new force compare with the old one? Is the force attractive or repulsive?

4. A physicist charges an object with a positively-charged rod. If the object develops a negative charge, how did the physicist charge the object?

5. If a circuit has a low voltage, is there any way to get a lot of energy from it? If so, how?

6. You are blindfolded and handed two extension cords. Both have the same current running through them. If the cord in your left hand is warmer than the one in your right hand, which cord is thicker?

7. Draw the conventional current flow in the following circuit with a dashed line. Draw the flow of electrons with a solid line.

8. In most chandeliers today, when one light bulb goes out, the rest stay lit. Are the light bulbs wired in series or in parallel?

9. The atoms in an object are not aligned. Is the object a magnet?

10. You cut a magnet in half - right between the north and south poles. How many north poles and south poles do you now have?

TEST FOR MODULE #13
(You may use the periodic chart to answer these questions.)

1. Define the following terms:

a. Nucleus
b. Atomic number
c. Mass number
d. Isotopes
e. Element
f. Radioactive isotope

2. Which is larger, an electron or a proton? Which is positively-charged?

3. What causes the strong nuclear force, and why can this force act only over very short distances?

4. List the number of protons, neutrons, and electrons in the following atoms:

a. ^{48}Ca   b. ^{124}Sn   c. ^{109}Ag

5. Draw an illustration of what the Bohr model says a ^{23}Na atom looks like.

6. Which of the following atoms are isotopes?

^{144}Ce, ^{144}Nd, ^{144}Sm, ^{145}Nd

7. A radioactive isotope has a half-life of 3 hours. If a scientist has 30 grams of the isotope, how much is left after 15 hours?

8. What is the daughter product in the beta decay of ^{144}Ce?

9. What is the daughter product in the alpha decay of ^{220}Rn?

10. What is the daughter product in the gamma decay of ^{239}U?

11. If a piece of paper is placed between a radioactive isotope and a person, which kind of radioactive particle will the person be protected from?

TEST FOR MODULE #14

1. Define the following terms:

a. Transverse wave
b. Longitudinal wave
c. Supersonic speed
d. Sonic boom
e. Pitch

2. A recorder is a woodwind instrument in which the player blows into a tube, setting up a wave in the tube. A musician has 2 recorders. The first one is rather short, and the second one is significantly longer. Which recorder is capable of playing notes with the lowest pitch?

3. A popular science fiction movie was advertised with the slogan, "In space, no one can hear you scream." Why is this a true statement?

4. Do sound waves oscillate parallel to or perpendicular to the direction in which the wave travels?

5. What is the speed of sound in air that has a temperature of 25 $^{\circ}$C?

6. A sound wave traveling through 17 $^{\circ}$C air has a wavelength of 2 meters. What is the frequency of the sound wave?

7. Which waves have the longest wavelength: sonic waves, infrasonic waves, or ultrasonic waves?

8. During a thunderstorm, the temperature is 10 $^{\circ}$C. If you see a lightning strike and then hear the thunder 2 seconds later, how far away did the lightning strike?

9. A ship blows its horn. Some of the sound waves travel through the air and then hit the water. Will the sound waves travel faster in the water or in the air?

10. A man and woman are singing a duet. The man sings the low notes and the woman sings the high notes. The woman, since she is singing the melody, is louder than the man. Are the wavelengths of the man's sound waves longer than, shorter than, or the same size as those of the woman? Is the frequency of the man's sound waves lower or higher than those of the woman? What about the amplitudes of the waves? What about the speed of the sound waves?

11. A jet travels through 10 $^{\circ}$C air at Mach 3. What is its speed in meters per second?

12. You are driving in a city that has a siren which sounds for 30 seconds every day at noon. You stop at a stoplight and then hear the sound of the siren. The stoplight then turns green, and you start driving. As you speed up, you notice that the pitch of the siren keeps getting higher.

Are you driving towards or away from the siren? (Assume that the true pitch of the siren stays constant.)

13. An amplifier takes a 30 decibel sound and turns it into an 80 decibel sound. How many times larger is the intensity of the sound waves coming out of the amplifier as compared to the intensity of the sound waves going into the amplifier?

TEST FOR MODULE #15

1. Define the following terms:

a. Electromagnetic wave
b. The Law of Reflection

2. Describe light according to the currently accepted theory. Be very detailed.

3. The speed of light in glass is 185,000,000 m/sec. If a particle is traveling through air at 200,000,000 m/sec, what has to happen to the particle's speed if it enters the glass?

4. Which has a higher frequency: green light or yellow light?

5. Which has longer wavelength: ultraviolet rays or radio waves?

6. Draw what happens to the light ray when it hits the right side of the aquarium.

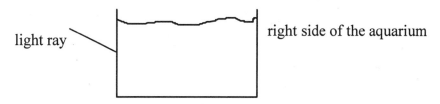

light ray   right side of the aquarium

7. You are on a diving board looking down into a pool. You see a quarter at the bottom of the pool, about three feet in front of you. Why is the quarter really not 3 feet in front of you?

8. You want to concentrate light coming from a weak light source by focusing it all on a single point. Would you use a converging or diverging lens to do this?

9. What does the eye do to change its focus?

For problems 10-12, remember that blue + green = cyan, red + blue = magenta, and red + green = yellow.

10. If a computer monitor makes a green dot and then puts a blue dot in the same place, what color will you see?

11. If cyan and magenta ink is mixed, what color ink will you get?

12. If you shine red light on a yellow shirt, what color will the shirt appear?

TEST FOR MODULE #16

1. Define the following terms:

a. Nuclear fusion
b. Nuclear fission
c. Critical mass
d. Star magnitude
e. Light year
f. Galaxy

2. What nuclear process occurs in the sun's core?

3. A ^{4}He nucleus and a ^{7}Li nucleus collide and form a ^{10}B nucleus and a neutron. Is this nuclear fusion or nuclear fission?

4. A scientist studies a process in which a neutron strikes a ^{216}Pu nucleus to make a ^{104}Cd nucleus, a ^{110}Pd nucleus, and three neutrons. If the scientist measures the mass of the ^{216}Pu and the original neutron and then subtracts the mass of the ^{104}Cd nucleus, the mass of the ^{110}Pd nucleus, and the mass of the three neutrons, will the scientist get a positive number, a negative number, or zero?

5. Is it possible for a nuclear power plant to experience a nuclear explosion? Why or why not?

6. Using the H-R diagram below, classify the following stars:

a. Magnitude 5, Spectral Letter G
c. Magnitude 11, Spectral Letter A

b. Magnitude 0, Spectral Letter M
d. Magnitude -5, Spectral Letter G

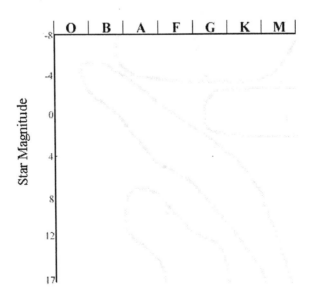

7. Which is the brightest of the stars in problem #6?

8. Which is the hottest of the stars in problem #6?

9. What are the three types of variable stars?

10. What are Cepheid variables and why are they important in astronomy?

11. What is the name of the galaxy to which earth's solar system belongs? What type of galaxy is it?

12. What is the red shift?

Solutions To The

Tests

ANSWERS TO THE TEST FOR MODULE #1

1. a. <u>Atom</u> - The smallest stable unit of matter in Creation

b. <u>Molecule</u> - Two or more atoms linked together to make a substance with unique properties

c. <u>Concentration</u> - The quantity of a substance within a certain volume of space

2. <u>Sulfur dioxide is made up of molecules.</u> We can tell this because it does not have the properties of sulfur (a yellow powder) but we are told that it contains sulfur atoms. The only way an atom can give up its properties is to become a part of a molecule.

3. <u>The statue will have a copper color underneath the green powder.</u> We learned from Experiment 1.1 and the subsequent discussion that the green powder we see on statues and the like is copper hydroxycarbonate. This molecule needs copper to form. Thus, the statue must have copper on it in order for the copper hydroxycarbonate to have formed on it.

4. <u>Picture A represents a bunch of molecules, whereas picture B is a representation of a bunch of atoms.</u>

5. <u>This is a measurement of mass.</u> The metric unit gram measures mass, regardless of the prefix in front of it.

6. <u>The prefix "kilo" is used to represent the number "1,000."</u>

7. First, we convert the number to a fractional form:

$$\frac{1.6 \text{ m}}{1}$$

Next, since we want to convert from meters to centimeters, we need to remember that "centi" means "0.01." So one centimeter is the same thing as 0.01 meters. Thus:

$$1 \text{ cm} = 0.01 \text{ m}$$

That's our conversion relationship. Since we want to end up with cm in the end, then we must multiply the measurement by a fraction that has meters on the bottom (to cancel the meter unit that is there) and cm on the top (so that cm is the unit we are left with). Remember, the numbers next to the units in the relationship above go with the units. Thus, since "m" goes on the bottom of the fraction, so does "0.01." Since "cm" goes on the top, so does "1."

$$\frac{1.6 \text{ m}}{1} \times \frac{1 \text{ cm}}{0.01 \text{ m}} = 160 \text{ cm}$$

Therefore, 1.6 m is the same as <u>160 cm</u>.

8. First, we convert the number to a fractional form:

$$\frac{0.12 \text{ kL}}{1}$$

Next, since we want to convert from kiloliters to L, we need to remember that "kilo" means "1,000." So one kiloliter is the same thing as 1,000 liters. Thus:

$$1 \text{ kL} = 1,000 \text{ L}$$

That's our conversion relationship. Since we want to end up with L in the end, then we must multiply the measurement by a fraction that has kiloliters on the bottom (to cancel the kL unit that is there) and L on the top (so that L is the unit we are left with):

$$\frac{0.12 \text{ \sout{kL}}}{1} \times \frac{1,000 \text{ L}}{1 \text{ \sout{kL}}} = 120 \text{ L}$$

Thus, 0.12 kL is the same as <u>120 L</u>.

9. $$\frac{45.1 \text{ \sout{kg}}}{1} \times \frac{1 \text{ slug}}{14.59 \text{ \sout{kg}}} = 3.09 \text{ slugs}$$

There are <u>3.09 slugs</u> in 45.1 kg. Note that I rounded the answer. The real answer was "3.091158328," but there are simply too many digits in that number. When you take chemistry, you will learn about significant figures, a concept that tells you where to round numbers off. For right now, don't worry about it. If you rounded at a different spot than I did, that's fine.

10. <u>The second bucket has the more powerful cleaner.</u> Remember, concentration is what is important. Both buckets have the same amount of ammonia in them, but the ammonia in the second bucket is contained in a smaller volume. Thus, the second bucket is more concentrated in ammonia.

ANSWERS TO THE TEST FOR MODULE #2

1. a. <u>Humidity</u> - The moisture content of air

b. <u>Absolute humidity</u> - The mass of water vapor contained in a certain volume of air

c. <u>Relative humidity</u> - A quantity expressing humidity as a percentage of the maximum absolute humidity for that particular temperature

d. <u>Greenhouse effect</u> - The process by which certain gases (principally water, carbon dioxide, and methane) trap heat that would otherwise escape the earth and radiate into space

e. <u>Parts per million</u> - The number of molecules (or atoms) of a substance in a mixture for every one million molecules (or atoms) in that mixture

2. <u>Water evaporates slowly under conditions of high humidity</u>. Remember, the higher the humidity, the more water vapor is already in the air. This makes it harder to put more water vapor in the air, which is what evaporation does.

3. <u>No, sweating will not cool you down</u>. When the humidity is 100%, water will not evaporate, and that is what gives sweat its cooling effect.

4. <u>Nitrogen makes up the majority of the air we inhale</u>. See Figure 2.3.

5. <u>Nitrogen makes up the majority of the air we exhale</u>. See Figure 2.3.

6. <u>The substance will burn the fastest in the second trial</u>. In that trial, the oxygen concentration is more than twice that of the first trial. After all, the room's air supply would have been about 21% oxygen. Since the second trial used an air mixture that was 50% oxygen, the larger concentration of air would result in faster combustion.

7. <u>Ozone blocks the ultraviolet light from the sun</u>. Without it, life could not exist on the planet.

8. <u>Ground-level ozone concentrations should be decreased</u>. Remember, ozone is a poison. We do not want to breathe it. We want it all up in the ozone layer.

9. <u>Global warming is not happening today</u>. See Figure 2.4.

10. Remember, we know the relationship between percent and ppm, so we can convert using the factor-label method.

$$\frac{1\%}{1} \times \frac{10,000 \text{ ppm}}{1\%} = 10,000 \text{ ppm}$$

A concentration of 1% is the same as <u>10,000 ppm</u>.

11. Remember, we know the relationship between ppm and percent. We can therefore just use the factor-label method to figure out the answer.

$$\frac{0.019 \text{ ppm}}{1} \times \frac{1\%}{10,000 \text{ ppm}} = 0.0000019\%$$

A concentration of 0.019 ppm is equal to <u>0.0000019 %</u>.

12. <u>Catalytic converters reduced the concentration of carbon monoxide</u>.

ANSWERS TO THE TEST FOR MODULE #3

1. a. <u>Atmosphere</u> - The mass of air surrounding a planet

b. <u>Barometer</u> - An instrument used to measure atmospheric pressure

c. <u>Homosphere</u> - The lower layer of earth's atmosphere, which exists from ground level to roughly 80 kilometers (50 miles) above sea level

d. <u>Heterosphere</u> - The upper layer of earth's atmosphere, which exists higher than 80 kilometers (50 miles) above sea level

e. <u>Jet streams</u> - A narrow band of high-speed winds that circle the earth, blowing from west to east

f. <u>Heat</u> - Energy that is being transferred

g. <u>Temperature</u> - A measure of the energy of motion in a substance's molecules

2. <u>You would study the troposphere</u> because that's where the majority of weather phenomena are.

3. <u>You would study the stratosphere</u> because that's where the ozone layer is.

4. <u>It must have come from the heterosphere.</u> Air in the homosphere is all 78% nitrogen, 21% oxygen, and 1% other.

5. <u>The troposphere, stratosphere, and mesosphere are in the homosphere.</u>

6. <u>The thermosphere and exosphere are in the heterosphere.</u>

7. <u>The height of the column will decrease.</u> Remember, there is a column of liquid there in the first place because of an imbalance between air pressure inside and outside of the column. As air seeps in, that imbalance will decrease, resulting in a smaller column. If the leak is big enough so that as much air can get in as possible, the level of liquid inside and outside the column will be the same.

8. <u>Temperature increases with increasing altitude in the stratosphere</u> because of the ozone layer.

9. <u>The "ozone hole" is a seasonal phenomenon located only at the South Pole because ozone cannot be depleted by CFCs without the aid of the Polar Vortex.</u> Since the Polar Vortex is seasonal and only exists at the South Pole, the same can be said for the "ozone hole."

10. <u>Heat is energy that is being transferred. To freeze water, energy must be transferred from the water to the surroundings.</u> Thus, water freezes because of heat!

11. <u>The average speed of the molecules in the air would decrease</u>. Remember, temperature measures the average motional energy of molecules, which is directly related to their speed. Since temperature decreases with increasing altitude in the troposphere, then the average energy of the molecules in the troposphere decreases, which means their speeds decrease as well.

ANSWERS TO THE TEST FOR MODULE #4

1. a. <u>Electrolysis</u> - Using electricity to break a molecule down into its constituent elements

b. <u>Polar molecule</u> - A molecule that has slight positive and negative charges due to an imbalance in the way electrons are shared

c. <u>Solvent</u> - A liquid substance capable of dissolving other substances

d. <u>Solute</u> - A substance that is dissolved in a solvent

e. <u>Cohesion</u> - The phenomenon that occurs when individual molecules are so strongly attracted to each other that they tend to stay together, even when exposed to tension

f. <u>Hard water</u> - Water that has certain dissolved ions in it, predominately calcium ions

2. <u>H_2O</u>

3. <u>H_4O would be the more likely erroneous result</u>. If the oxygen was being absorbed by the metal covering on the battery, then there would be less collected in the tube. This would make it look like water molecules had a lot *less* oxygen. HO_2 indicates *more* oxygen.

4. <u>Hydrogen bonding</u> keeps the water molecules close together.

5. <u>There are three atoms in a carbon dioxide molecule and 2 atoms in a carbon monoxide molecule</u>. Remember, each capital letter signifies an atom, and if there is no subscript, that means there is only one atom.

6. The chemical formula is <u>C_8H_{18}</u>.

7. <u>Water molecules are polar because both the oxygen atoms and the hydrogen atoms are fighting over the electrons they are supposed to be sharing. Oxygen can pull on those electrons harder, so it gets more than its fair share of electrons, making it slightly negative. Since the hydrogen atoms get less than their fair share of electrons, they end up slightly positive.</u>

8. <u>It is nonpolar</u> because water dissolves both polar and ionic substances.

9. <u>It will not dissolve in vegetable oil</u> because if it dissolves in water, it is either ionic or polar. Either way, such a substance will not dissolve in a nonpolar liquid like vegetable oil.

10. <u>No</u>. Hard water is the result of the calcium-containing compounds in the region from which the water is taken.

ANSWERS TO THE TEST FOR MODULE #5

1. a. <u>Transpiration</u> - Emission of water vapor from plants

b. <u>Condensation</u> - The process by which water vapor turns into liquid water

c. <u>Residence time</u> - The average time a given molecule of water will stay in a given water source

d. <u>Percolation</u> - The process by which water passes from above the water table to below it

e. <u>Adiabatic cooling</u> - The cooling of a gas that happens when the gas expands

2. It resides in the <u>oceans</u>.

3. <u>Groundwater</u> is the largest source of liquid freshwater.

4. It is in the <u>atmosphere</u>. Remember, transpiration takes water from plants and puts it in the atmosphere.

5. <u>Evaporation</u> allowed it to leave the ocean and <u>condensation</u> put it in the cloud. The student needs both, but partial credit can be given for one.

6. The residence time is longest where the least amount of water exchange takes place. Since evaporation is the only way out of the ocean, it will take a long time for a molecule of water to leave the ocean. Thus, <u>the ocean</u> has a longer residence time.

7. <u>It contains saltwater</u>. If the only means of losing water is evaporation, then the salts continue to concentrate, making saltwater.

8. <u>They tell us that the earth can't be billions of years old</u>. If the earth was more than a million years old, the oceans would be much saltier.

9. <u>It does not form an iceberg</u>. Icebergs are freshwater and come from glaciers. Sea ice has salt mixed in with it.

10. <u>Precipitation</u> is responsible for glaciers. Remember, glaciers start because of snow, and snow is precipitation.

11. <u>Adiabatic cooling</u> causes the condensation that makes clouds. If the student talks about particles as cloud condensation nuclei, give partial credit, but condensation won't happen without cooling!

12. <u>The temperature will increase</u>. Gases cool as they expand and heat up as they are compressed.

13. <u>The depth of the water table will decrease</u>. If a lot more water enters the soil, then more soil than usual will become saturated. That means you don't have to go down as far to find the saturated water, so the water table is higher.

14. <u>The nature of groundwater flow makes it such that a lake can be polluted by groundwater that originally soaked into the soil hundreds of miles away</u>. When you find a lake polluted by groundwater pollution, how will you know where it came from?

ANSWERS TO THE TEST FOR MODULE #6

1. a. <u>Sedimentary rock</u> - Rock formed when heat, pressure, and chemical reactions cement sediments together

b. <u>Plastic rock</u> - Rock that behaves like something between a liquid and a solid

c. <u>Fault</u> - The boundary between a section of moving rock and a section of stationary rock

d. <u>Epicenter</u> - The point on the surface of the earth directly above an earthquake's focus

2. a. <u>lithosphere</u>
   b. <u>mantle</u>
   c. <u>outer core</u>
   d. <u>inner core</u>

3. <u>seismic waves</u>

4. The Moho separates the lithosphere from the mantle. It is therefore <u>between (a) and (b)</u>.

5. <u>Electrical flow in the core</u> causes the earth's magnetic field.

6. <u>The dynamo theory and the rapid decay theory</u> both attempt to explain the earth's magnetic field. <u>The rapid decay theory is more scientifically valid.</u>

7. <u>The earth's magnetic field blocks cosmic rays from the sun.</u>

8. <u>The trench is probably the site where one plate slides under another.</u>

9. <u>The region nearest the plate boundary should have more earthquakes.</u>

10. Each step on the Richter scale means a factor of 32 in energy. Since the quake and aftershock are off by 2 units, the quake was 32x32 = <u>1,024 times more energetic than the aftershock.</u>

11. <u>You expect to find both volcanic mountains and domed mountains.</u>

12. <u>You expect to find fault-block mountains.</u>

ANSWERS TO THE TEST FOR MODULE #7

1. a. <u>Aphelion</u> - The point at which the earth is farthest from the sun

b. <u>Perihelion</u> - The point at which the earth is closest to the sun

c. <u>Coriolis effect</u> - The way in which the rotation of the earth bends the path of winds, sea currents, and objects that fly through different latitudes

d. <u>Air mass</u> - A large body of air with relatively uniform pressure, temperature, and humidity

e. <u>Weather front</u> - A boundary between two air masses

2. a. <u>stratus</u> (stratonimbus is okay also) b. <u>cirrus</u> c. <u>cumulus</u> (cumulonimbus is okay also)

3. The three main factors are thermal energy, uneven distribution of thermal energy, and water vapor in the atmosphere. Winds are caused by imbalances of temperature, which is due to <u>uneven distribution of thermal energy</u>.

4. Dark clouds have a "nimbo" prefix or a "nimbus" suffix. The proper term is <u>cumulonimbus</u>, but nimbocumulus works as well.

5. Insolation stands for incoming solar radiation, which is sunlight. If a region gets a lot of sunlight, <u>it is warm</u>.

6. From the winter solstice to the summer solstice, day times in the Southern Hemisphere decrease because the Southern Hemisphere begins pointing away from the sun. At the spring equinox, the day length is 12 hours. Since the day time is <u>decreasing</u> after the winter solstice, then the day times must still be <u>greater than 12 hours</u> from the winter solstice to the spring equinox.

7. <u>The Northern Hemisphere is pointed away from the sun at perihelion</u>. This is a much greater effect than the distance from the sun.

8. <u>Life would cease to exist</u> because the difference in temperature between day and night would be too severe.

9. <u>The change in temperature caused by air changing latitude along with the Coriolis effect</u> cause the global wind patterns that we see on the earth.

10. <u>Local winds</u> interfere with the global wind patterns.

11. This is a <u>continental tropical</u> air mass.

12. This cloud progression is typical of a warm front, which causes <u>long, lighter rain</u>.

ANSWERS TO THE TEST FOR MODULE #8

1. a. <u>Updraft</u> - A current of rising air

b. <u>Insulator</u> - A substance that does not conduct electricity very well

2. <u>The Bergeron process</u> is at work here, because that's the process that says the precipitation starts out as snow and melts into rain as it reaches warm air.

3. <u>The dew point is the temperature at which water vapor condenses out of the air onto ground-level surfaces. It is influenced by humidity and pressure.</u>

4. <u>The thunderstorm is in its dissipation stage. No hail will exist</u> because hail must have an updraft to form.

5. <u>These thunderstorms are comprised of several cells,</u> each of which produces heavy rains for about 30 minutes or less.

6. <u>The charge imbalance that causes lightning starts in the cloud and cannot form unless the cloud is tall.</u> Remember, the charge imbalance comes from the millions of collisions that take place between the falling raindrop or ice crystal and other things in the cloud. Without a tall cloud, there will not be enough collisions to cause the charge imbalance.

7. <u>The stepped leader forms first.</u>

8. <u>Lightning causes thunder by heating up the air through which it passes. That heat generates a wave that we detect as sound.</u>

9. <u>Sheet lightning cannot strike a person</u> because it is cloud-to-cloud lightning.

10. <u>Yes, it has touched ground.</u> That marks the beginning of the organization stage.

11. <u>Wind speed</u> differentiates all of the classifications of potential hurricanes.

12. <u>The eye</u> is safe and calm.

13. <u>The pressure is the same</u> because they are on the same isobar.

14. <u>The pressure in Chicago is higher</u> because Chicago is more than 3 isobars from the "L," while New York is only between 2 and 3 isobars from the "L."

15. <u>New York might be experiencing thunderstorms</u> because of the cold front.

16. <u>Houston should expect warmer weather</u> since it is in the path of a warm front.

ANSWERS TO THE TEST FOR MODULE #9

1. a. <u>Reference point</u> - A point against which position is measured

b. <u>Vector quantity</u> - A physical measurement that contains directional information

c. <u>Scalar quantity</u> - A physical measurement that does not contain directional information

d. <u>Acceleration</u> - The time rate of change of an object's velocity

e. <u>Free fall</u> - The state of an object that is falling towards the earth with nothing inhibiting its fall

2. In order for motion to occur, an object's position must change. In order to determine position, there must be a reference point. <u>The reference point allows you to determine whether or not position changes</u>. Note: the statement "all motion is relative" deserves half credit.

3. a. <u>Your grandmother is in motion relative to you</u>. Even though your grandmother is standing still, her position relative to you is changing. Thus, she is in motion relative to you.

b. <u>Your mother is motionless relative to you</u>. Her position relative to you does not change. She is therefore motionless with respect to you.

4. This problem gives us distance and time and asks for speed. We know it is asking for speed because a distance unit divided by a time unit is speed or velocity. There is no direction here, so we are talking about speed. Thus, we need to use Equation (9.1). The problem wants the answer in miles per hour, however. We are given the time in minutes. Thus, we must make a conversion first:

$$\frac{30 \cancel{\text{minutes}}}{1} \times \frac{1 \text{ hour}}{60 \cancel{\text{minutes}}} = 0.5 \text{ hours}$$

Now we can use our speed equation:

$$\text{speed} = \frac{40 \text{ miles}}{0.5 \text{ hours}} = 80 \underline{\frac{\text{miles}}{\text{hour}}}$$

The driver was speeding!

5. The problem wants velocity, which is speed with a direction. To get speed, we will use Equation (9.1). Unfortunately, the problem tells us to give the answer in meters per second, but the distance is in km and the time is in minutes. Thus, we need to do two conversions:

$$\frac{4.1 \cancel{\text{minutes}}}{1} \times \frac{60 \text{ seconds}}{1 \cancel{\text{minute}}} = 246 \text{ seconds}$$

$$\frac{1 \ \text{km}}{1} \times \frac{1000 \ \text{m}}{1 \ \text{km}} = 1000 \, \text{m}$$

Now we can use our speed equation:

$$\text{speed} = \frac{1000 \ \text{m}}{246 \ \text{seconds}} = 4.07 \ \frac{\text{meters}}{\text{second}}$$

That's not quite the answer. The problem wants velocity, which includes direction. Thus, the answer is <u>4.07 meters/second west</u>.

6. a. This measurement has a distance unit divided by a time unit. That's speed or velocity. Since no direction is given, this is <u>speed</u>.

b. The unit of feet by itself measures <u>distance</u>.

c. This measurement has a distance unit divided by a time unit. That's speed or velocity. Since a direction is given, this is <u>velocity</u>.

d. This measurement has a distance unit divided by a time unit squared. That's <u>acceleration</u>. The direction is necessary because acceleration is a vector quantity.

7. As the picture shows, the eagle is behind the rabbit, but they are both traveling in the same direction. Thus, we get their relative velocity by subtracting their individual velocities:

relative velocity = 5.4 meters/second - 4.4 meters/second = 1.0 meter/second

Since the rabbit is traveling faster than the eagle, the rabbit is pulling away. Thus, the relative velocity is <u>1.0 meter per second away from each other</u>.

8. Since the velocity is not changing, <u>the acceleration is zero</u>.

9. The initial velocity is 12 meters per second north, and the final velocity is 0. The time is 3 seconds. This is a straightforward application of Equation (9.2).

$$\text{acceleration} = \frac{\text{final velocity} - \text{initial velocity}}{\text{time}}$$

$$\text{acceleration} = \frac{0 \ \frac{\text{meters}}{\text{second}} - 12 \ \frac{\text{meters}}{\text{second}}}{3 \ \text{seconds}} = \frac{-12 \ \frac{\text{meters}}{\text{second}}}{3 \ \text{seconds}} = -4 \ \frac{\text{meters}}{\text{second}^2}$$

The negative just tells us that the acceleration is in the opposite direction of velocity. Thus, the acceleration is <u>4 meters/second^2 south</u>.

10. This is another application of Equation (9.2), because we are given time (5 seconds), initial velocity (0) and final velocity (60 miles per hour). We can't use the equation yet, however, because our time units do not agree. We'll fix that first:

$$\frac{5 \ \text{seconds}}{1} \times \frac{1 \ \text{hour}}{3600 \ \text{seconds}} = 0.00139 \ \text{hours}$$

Now that we have all time units in agreement, we can use the acceleration equation:

$$\text{acceleration} = \frac{\text{final velocity} - \text{initial velocity}}{\text{time}}$$

$$\text{acceleration} = \frac{60 \ \frac{\text{miles}}{\text{hour}} - 0 \ \frac{\text{miles}}{\text{hour}}}{0.00139 \ \text{hours}} = \frac{60 \ \frac{\text{miles}}{\text{hour}}}{0.00139 \ \text{hours}} = 43165 \ \frac{\text{miles}}{\text{hour}^2}$$

The car speeds up, so acceleration is in the same direction as velocity. The answer, then, is 43165 miles/hour² north.

11. Neither has greater acceleration. Both objects are falling near the surface of the earth; thus, they are each in free fall. That means they both have equal acceleration. The ball was given more *initial velocity*, so it will travel faster. The acceleration on both is the same, however.

12. The feather is more affected by air resistance than the rock. This is the same situation as Experiment 9.2.

13. The rock is in free fall, so we can use Equation (9.3). Since the problem wants the answer in feet, we need to use 32 feet per second² as the acceleration.

$$\text{distance} = \frac{1}{2} \cdot (\text{acceleration}) \cdot (\text{time})^2$$

$$\text{distance} = \frac{1}{2} \cdot (32 \ \frac{\text{feet}}{\text{second}^2}) \cdot (1.3 \ \text{seconds})^2$$

$$\text{distance} = \frac{1}{2} \cdot (32 \ \frac{\text{feet}}{\text{second}^2}) \cdot (1.69 \ \text{second}^2) = 27.04 \ \text{feet}$$

ANSWERS TO THE TEST FOR MODULE #10

1. a. <u>Friction</u> - A force resulting from the contact of two surfaces. This force opposes motion.

b. <u>Kinetic friction</u> - The friction that exists between surfaces when at least one of those surfaces is moving relative to the other

c. <u>Static friction</u> - The friction that exists between surfaces when neither surface is moving relative to the other

2. <u>Newton's First Law</u> - The velocity of an object will not change unless the object is acted on by an outside force.

<u>Newton's Second Law</u> - When an object is acted on by an outside force, the strength of that force is equal to the mass of the object times the resulting acceleration.

<u>Newton's Third Law</u> - For every action, there is an equal and opposite reaction.

3. <u>The pilot must drop the bombs before the plane reaches the airfield</u>. The bombs will have a velocity equal to that of the plane when they are dropped. Thus, they will continue to fly north as they fall. In order to hit the airfield, then, they must be dropped south of it.

4. <u>The mouse will start moving straight, skidding off of the disk</u>. This is like Experiment 10.2. When the disk stops, the mouse has a velocity pointed in a certain direction. Without sufficient time, the frictional force will not be able to keep the mouse on the disk. This will cause the mouse to start traveling in a straight line, in the direction it was moving right before the disk stopped.

5. Since we are ignoring friction, this is an easy problem:

$$F = (mass) \cdot (acceleration)$$

$$F = (1.0 \, kg) \cdot \left(4.0 \, \frac{m}{sec^2} \right) = 4.0 \text{ Newtons}$$

The cube is being pulled with a force of <u>4.0 Newtons down the tray</u>.

6. The baseball player is slowing, because his velocity is north but the acceleration is directed south. Friction slows things down. This is the only force in the problem, since nothing is pulling or pushing on the player. Thus, the force that results from the acceleration will be the frictional force.

$$F = (\text{mass}) \cdot (\text{acceleration})$$

$$F = (75\,\text{kg}) \cdot \left(5.0 \ \frac{\text{m}}{\text{sec}^2}\right) = 375 \ \text{Newtons}$$

The frictional force is 375 Newtons to the south.

7. If the cart is moving with a constant velocity, that means the acceleration (and thus the total force) equals zero. Thus, the man must be pushing with just enough force to counteract friction. Thus, the frictional force is 200 Newtons against the motion of the cart. This is kinetic friction, because the cart is moving.

8. Since static friction is always greater than kinetic friction, 11 Newtons refers to static friction and 8 Newtons refers to kinetic friction.

9. If the dog resists with 30 Newtons, that force adds to the static frictional force (20 Newtons), because friction also opposes motion. Thus, there are a total of 50 Newtons of force resisting motion. To get the dog moving, a force greater than 50 Newtons must be used. Once the dog is moving, the acceleration is 1.0 meters per second^2, so the total force is:

$$F = (\text{mass}) \cdot (\text{acceleration})$$

$$F = (30\,\text{kg}) \cdot \left(1.0 \ \frac{\text{m}}{\text{sec}^2}\right) = 30 \ \text{Newtons}$$

Once the dog is moving, it still resists with 30 Newtons of force, and the kinetic frictional force resists with 7 Newtons of force, for a total of 37 Newtons. That force subtracts from the force the woman applies, because it resists motion. Thus, to get a total of 30 Newtons, the woman must pull with a force of 67 Newtons to the west.

10. The ball exerts a force on the bat. This is the equal and opposite force demanded by Newton's Third Law.

ANSWERS TO THE TEST FOR MODULE #11

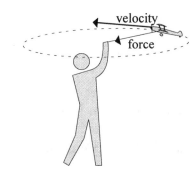

1.

2. <u>The weakest of the fundamental forces is gravity</u>.

3. <u>The ball exerts an equal and opposite force on the earth</u>.

4. When the 5 kg mass is replaced by a 20 kg mass, the mass is increased by a factor of 4. This increases the gravitational force by a factor of 4. The 2 kg mass is replaced by a 12 kg mass, increasing that mass by a factor of 6. Thus, the total change is 4x6 = 24. <u>The new gravitational force, then, is 24 times larger than the old one</u>.

5. The only difference is that the distance between the objects was increased by a factor of 3. The gravitational force decreases when the distance between the objects increases. It is decreased according to the square of that increase. Thus, the change is $3^2 = 9$. <u>The new gravitational force, then, is 9 times smaller than the old one</u>.

6. The first one moves faster, because it takes less time to make the orbit. The principles of circular motion state that the faster the speed, the greater the centripetal force necessary. Gravity supplies the centripetal force in an orbit. Thus, <u>the first moon experiences the greatest gravitational force</u>.

7. <u>Jupiter, Saturn, Uranus, Neptune, Pluto</u>

8. The planet farthest away from the sun receives the least insolation. Thus, the answer is <u>Pluto</u>.

9. Comet orbits are highly elliptical and pass close to the sun. Thus, the answer is <u>c</u>.

10. <u>The coma and tail come and go</u>, depending on the proximity to the sun.

11. <u>The Kuiper belt is a source of short-term comets</u>. An answer of just "comets" receives only half credit.

12. <u>The graviton theory</u> states that gravity is caused by the exchange of particles.

13. <u>The General Theory of Relativity</u> would be true.

ANSWERS TO THE TEST FOR MODULE #12

1 a. <u>Photons</u> - Small "packages" of light that act just like small particles

b. <u>Charging by conduction</u> - Charging an object by allowing it to come into contact with an object which already has an electrical charge

c. <u>Charging by induction</u> - Charging an object by forcing some of the charges to leave the object

d. <u>Electrical current</u> - The amount of charge that travels through an electrical circuit each second

e. <u>Conventional current</u> - Current that flows from the positive side of the battery to the negative side. This is the way current is drawn in circuit diagrams, even though it is wrong.

f. <u>Resistance</u> - A measure of how much a metal impedes the flow of electrons

g. <u>Open circuit</u> - A circuit that does not have a complete connection between the two sides of the battery. As a result, current does not flow.

2. Opposite charges attract one another. Thus, they will each pull the other to themselves.

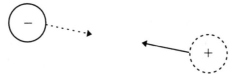

3. The electromagnetic force varies inversely with the square of the distance between magnets. Thus, if the distance is doubled, <u>the force decreases by a factor of four</u>. Since the poles are the same, this is a <u>repulsive</u> force.

4. The object developed a charge opposite of the rod. Thus, <u>the object was charged by induction</u>.

5. A low voltage means that each electron will have only a little bit of energy. <u>You can still get a lot of energy out of the circuit, however, as long as you put a lot of current through it.</u>

6. <u>The cord in your right hand is thicker.</u> The thicker the cord, the lower the resistance. The lower the resistance, the less heat will be made.

7. Conventional current runs from the positive end of the battery (the large line) to the negative end (the small line). The electron flow is opposite of that.

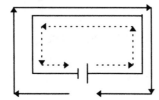

8. <u>The light bulbs are wired in parallel</u>.

9. <u>The object is not a magnet</u>.

10. <u>You will have two north poles and two south poles</u>. As soon as the magnet is cut, each pole will change. Since it is impossible to have just one pole of a magnet, the north pole will change into a magnet with a north and south pole. The south pole will also change into a magnet with a north and south pole. Thus, you will have 2 magnets, each with a north and south pole.

ANSWERS TO THE TEST FOR MODULE #13

1. a. <u>Nucleus</u> - The center of an atom, containing the protons and neutrons

b. <u>Atomic number</u> - The number of protons in an atom

c. <u>Mass number</u> - The sum of the number of neutrons and protons in the nucleus of an atom

d. <u>Isotopes</u> - Two or more atoms that have the same number of protons but different numbers of neutrons

e. <u>Element</u> - A collection of atoms that all have the same number of protons

f. <u>Radioactive isotope</u> - An atom whose nucleus is not stable

2. <u>A proton is larger than an electron. A proton is positively-charged.</u>

3. <u>The strong nuclear force is caused by the exchange of pions. It is a short-range force because pions can only exist for a short time.</u>

4. a. Since the chemical symbol is Ca, we can use the chart to learn that the atom has <u>20 protons</u>. This tells us there are also <u>20 electrons</u>. The mass number is the sum of protons and neutrons in the nucleus. Thus, there are <u>28 neutrons</u>.

b. Since the chemical symbol is Sn, we can use the chart to learn that the atom has <u>50 protons</u>. This tells us there are also <u>50 electrons</u>. The mass number is the sum or protons and neutrons in the nucleus. Thus, there are <u>74 neutrons</u>.

c. Since the chemical symbol is Ag, we can use the chart to learn that the atom has <u>47 protons</u>. This tells us there are also <u>47 electrons</u>. The mass number is the sum of protons and neutrons in the nucleus. Thus, there are <u>62 neutrons</u>.

5. All atoms symbolized with "Na" have 11 protons according to the chart. This also means there are 11 electrons. Two of them can go into the first Bohr orbit, and 8 can go in the second Bohr orbit. We will have to put the remaining one in the third Bohr orbit. The mass number indicates that there are 12 neutrons:

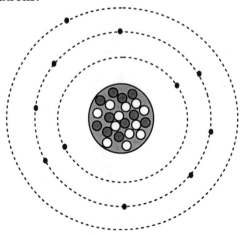

6. Isotopes have the same number of protons but different numbers of neutrons. Since chemical symbols tell us the number of protons, we are looking for atoms with the same chemical symbols but different mass numbers. Thus, ^{144}Nd and ^{145}Nd are isotopes.

7. In the first three hours, the 30 gram sample will be reduced to 15 grams. In the next three hours, it will be reduced to 7.5 grams. In the next 3 hours, it will be reduced to 3.75 grams. In the next three hours, it will be reduced to 1.875 grams. In the next 3 hours, it will be reduced to 0.9375 grams.

8. ^{144}Ce has 58 protons according to the chart. This means there must be 86 neutrons. In beta decay, a neutron turns into a proton. This will result in an atom with 59 protons and 85 neutrons, or ^{144}Pr.

9. ^{220}Rn has 86 protons according to the chart. This means there must 134 neutrons. In alpha decay, the nucleus loses 2 protons and 2 neutrons. This will result in an atom with 84 protons and 132 neutrons, or ^{216}Po.

10. In gamma decay, the number of protons and neutrons is unaffected by the decay. Thus, the daughter product is the same as the original ^{239}U.

11. The person would only be protected from alpha particles.

ANSWERS TO THE TEST FOR MODULE #14

1. a. <u>Transverse wave</u> - A wave whose propagation is perpendicular to its oscillation

b. <u>Longitudinal wave</u> - A wave whose propagation is parallel to its oscillation

c. <u>Supersonic speed</u> - Any speed that is faster than the speed of sound in the substance of interest

d. <u>Sonic boom</u> - The sound produced as a result of aircraft traveling at or above Mach 1

e. <u>Pitch</u> - The highness or lowness of a sound

2. Low pitch means low frequency. Since wavelength and frequency are inversely proportional, low frequency means large wavelength. Thus, the <u>longer recorder produces the lower notes.</u>

3. <u>No one can hear you scream in space because there is little to no air in space. As a result, sound waves cannot travel,</u> because there is no medium for them to oscillate.

4. Sound waves are longitudinal waves. This means <u>they oscillate parallel to the direction in which they travel.</u>

5. The speed of sound is given by Equation (14.2):

$$v = (331.5 + 0.60 \cdot T) \ \frac{m}{sec}$$

$$v = (331.5 + 0.60 \cdot 25) \frac{m}{sec} = \underline{346.5 \ \frac{m}{sec}}$$

6. To determine frequency, we need to use Equation (14.1). To use that equation, however, we need to know the speed and wavelength. We are given the wavelength (2 m), and we can get the speed from the temperature and Equation (14.2):

$$v = (331.5 + 0.60 \cdot T) \ \frac{m}{sec}$$

$$v = (331.5 + 0.60 \cdot 17) \frac{m}{sec} = 341.7 \ \frac{m}{sec}$$

Now that we have the speed, we can finally use Equation (14.1):

$$f = \frac{v}{\lambda}$$

$$f = \dfrac{341.7 \ \dfrac{\cancel{m}}{sec}}{2 \ \cancel{m}} = 170.85 \ \dfrac{1}{sec}$$

The frequency is <u>170.85 Hz</u>.

7. Infrasonic waves have the lowest frequencies, sonic waves have higher frequencies, and ultrasonic waves have the highest frequencies. Since frequency and wavelength are inversely proportional, <u>infrasonic waves have the longest wavelengths</u>.

8. To determine the distance, we will use the time difference between the lightning flash and the sound. We will assume that the light from the lightning reaches your eyes essentially at the same time as the lightning was formed. Thus, the time it takes for the sound to travel to you will determine the distance. First, then, we need to know the speed of sound:

$$v = (331.5 + 0.60 \cdot 10) \ \dfrac{m}{sec}$$

$$v = 337.5 \ \dfrac{m}{sec}$$

Now we can use Equation (14.3):

$$distance = (speed) \times (time)$$

$$distance = (337.5 \ \dfrac{m}{\cancel{sec}}) \times (2 \ \cancel{sec}) = \underline{675 \, m}$$

9. Sound waves travel faster in liquids than they do in gases. Thus, <u>the sound will travel faster in water</u>.

10. Since the man is singing the low notes, <u>the man's sound waves have the lower frequencies and the longer wavelengths</u>. The speed of sound depends only on the medium and temperature. Thus, <u>the speed of the man's sound waves are the same as that of the woman's sound waves</u>. Finally, amplitude determines loudness. Therefore, <u>the amplitude of the man's sound waves is smaller</u>.

11. To determine the speed of the jet, we first have to determine the speed of sound. After all, Mach 3 means 3 times the speed of sound. Thus, we need to know the speed of sound in order to determine the speed of the jet.

$$v = (331.5 + 0.60 \cdot T) \ \dfrac{m}{sec}$$

$$v = (331.5 + 0.60 \cdot 10)\frac{m}{sec} = 337.5 \frac{m}{sec}$$

Since sound travels at 337.5 m/sec, Mach 3 is 3 x (337.5 m/sec) = 1,012.5 m/sec.

12. Since the pitch of the siren gets higher, you are moving so that the wavelength of the sound waves seems smaller. This means you are moving towards the siren, because that will cause you to encounter the crests of the sound waves faster than you would if you were still.

13. The bel scale states that every bel unit corresponds to a factor of ten in the intensity of the sound waves. Thus, we need to determine how many bel units the sound was before and after the amplifier:

$$\frac{30 \text{ decibels}}{1} \times \frac{1 \text{ bel}}{10 \text{ decibels}} = 3 \text{ bels}$$

$$\frac{80 \text{ decibels}}{1} \times \frac{1 \text{ bel}}{10 \text{ decibels}} = 8 \text{ bels}$$

Since the sound is 5 bels louder after the amplifier, the increase in sound wave intensity is 5 factors of ten higher. Thus, the sound wave intensity is 10 x 10 x 10 x 10 x 10 = 100,000 times larger after the amplifier as compared to before.

ANSWERS TO THE TEST FOR MODULE #15

1. a. <u>Electromagnetic wave</u> - A transverse wave composed of an oscillating electrical field and a magnetic field that oscillates perpendicular to the electrical field

b. <u>The Law of Reflection</u> - The angle of reflection equals the angle of incidence

2. <u>Light is made up of individual packets. Each packet is made up of two perpendicular transverse waves: one which consists of an oscillating electrical field and one that is made up of an oscillating magnetic field.</u> Just the mention of both a particle and a wave is worth half credit, but the electrical and magnetic fields must be mentioned for full credit.

3. According to Einstein's Special Theory of Relativity, the speed of light in any substance is the maximum speed allowed. Thus, <u>the particle will have to slow down</u>.

4. Yellow light has a longer wavelength than green light. Thus, <u>green light has a higher frequency</u>.

5. <u>Radio waves have a longer wavelength</u>.

6. The light ray will bend towards the perpendicular when it refracts into the aquarium. There will also be a reflected ray there, but that does not affect the answer to this problem. When it reaches the other side, part of it will refract out and bend away from the perpendicular. The rest will reflect away. To get full credit, the student must show the ray leaving the aquarium and bending away from the perpendicular, and the student must also show the reflected ray with an angle equal to that of the incident ray. Count this problem as 4 points: one for the refracted ray, one for making it bend away from the perpendicular, one for the reflected ray, and one for the proper angle of reflection.

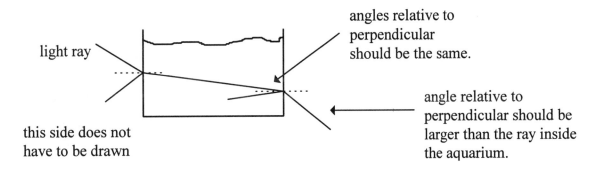

7. The quarter is not really 3 feet in front of you because <u>as the light from the quarter leaves the pool water, it bends. This causes a false image</u>. The quarter is really a bit closer to you.

8. You would use a <u>converging lens</u>. A diverging lens would cause the light rays to move away from each other, not converge to a single point.

9. To change its focus, the eye <u>changes the shape of the lens</u>.

For problems 10-12, remember that blue + green = cyan, red + blue = magenta, and red + green = yellow.

10. Since a computer screen shines light at your eyes, the colors add. Thus, you get a <u>cyan</u> dot.

11. Cyan ink absorbs all colors except blue and green. Thus, when white light shines on a cyan spot, only blue and green light reflect off of it. That's what gives it the cyan color. In the same way, magenta ink absorbs all colors except blue and red. Therefore, a mixture of the two will absorb all colors but blue. As a result, <u>the ink will be blue</u>.

12. Yellow absorbs all colors except red and green. Thus, the red light will be reflected back. That will be the *only* color reflected back, so it will appear <u>red</u>.

ANSWERS TO THE TEST FOR MODULE #16

1. a. <u>Nuclear fusion</u> - The process by which two or more small nuclei fuse to make a bigger nucleus

 b. <u>Nuclear fission</u> - The process by which a large nucleus is split into two smaller nuclei

c. <u>Critical mass</u> - The amount of isotope necessary to cause a chain reaction

d. <u>Star magnitude</u> - The brightness of a star on a scale of -8 to +17. The *smaller* the number, the *brighter* the star.

e. <u>Light year</u> - The distance light travels in one year

f. <u>Galaxy</u> - A massive ensemble of hundreds of millions of stars, all interacting through the gravitational force, orbiting around a common center

2. <u>Nuclear fusion occurs in the sun's core.</u>

3. <u>This is nuclear fusion</u>, since two small nuclei fused to make a bigger one.

4. <u>He would get a positive number.</u> In nuclear fission and fusion, mass is converted into energy. Thus, the mass of the materials produced will always be less than the mass of the starting materials.

5. <u>It is not possible, because a nuclear power plant does not have a critical mass of the isotope being used in nuclear fission.</u>

6. Using the magnitudes and spectral letters given, the stars fall on the H-R diagram as follows:

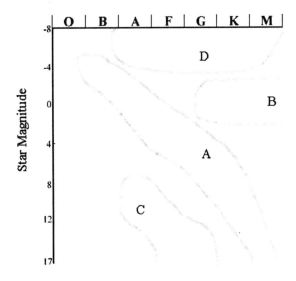

a. <u>Main sequence</u>   b. <u>Red giant</u>   c. <u>White dwarf</u>   d. <u>Supergiant</u>

7. The brightest star is the one with the lowest magnitude. Thus, (d) is the brightest star.

8. The hottest star is the one that is the farthest to the left on the H-R diagram, because temperature increases the farther left you travel on the H-R diagram. Thus, (c) is the hottest star.

9. The three types of variable stars are novas, supernovas, and pulsating variables.

10. Cepheid variables are variable stars whose magnitude and period have a direct relationship. They are important in astronomy because that relationship can be used to measure long distances in the universe.

11. Earth's solar system belongs to the Milky Way galaxy, which is a spiral galaxy.

12. The red shift is the phenomenon in which light that comes to the earth from other galaxies ends up having longer wavelengths than it should. Most astronomers interpret that as a Doppler shift resulting from the fact that the galaxies are moving away from us.